# The Gravity Guide

## Unveiling the Universe's Hidden Force

# The Gravity Guide

## Unveiling the Universe's Hidden Force

Pen Name: **PYURA**

**ANSHUMAN**

**2025 Edition**
**Pyura Books & Research**
www.pyurabooks.com

Pyura Books & Research may be purchased for educational, business or sales use. For information and communication write to publish@machaanlabs.com

Book cover designed by Sadia Akhtar

ISBN   979-8-86-389753-0

∞∞∞∞∞∞∞∞∞∞∞∞∞∞∞∞∞∞∞∞∞∞∞∞∞∞∞∞∞∞∞∞

WHERE NATURE, SCIENCE AND SOUL COLLIDE. 📖📕

# PYURA

For Aayra — with love and wonder.

And for every curious mind: students, teachers, lifelong learners, and seekers of truth across all ages.
This book is for those who pause to ask *why* an apple falls, *how* the planets move, or *what* holds the cosmos together.

Whether you're in a classroom or a quiet corner of the world, **The Gravity Guide** is your invitation to explore and experiment with the invisible force that shapes our universe, and to see the world a little differently each time you look up.

———————————————————

Profits from this book will be used for research and prototype development purposes.

# Table of Contents

# A Letter from the Author

Dear Reader,

There are forces in life that shape us long before we understand them. Some are emotional, some spiritual. But one—silent, invisible, yet constant—is gravity. Long before I studied its equations or followed its curvature through Einstein's spacetime, I felt it in the everyday poetry of falling leaves, tides that sighed on distant shores, and the way our feet cling lovingly to Earth.

This book was born not just from scientific curiosity, but from wonder. Wonder that grew into questions, and questions that demanded both rigour and reverence. It is my attempt to bridge worlds: of science and storytelling, mathematical concepts and intuition, fact and feeling.

You'll find in these pages the rigorous beauty of physics and the elegant mysteries of the cosmos, but also echoes of mythology, curiosity from childhood stargazing, and a hope that knowledge can still stir awe. Whether you're a student, teacher, dreamer, or seasoned explorer of the stars, I invite you to travel with me through Newton's orchard, Einstein's curvature, and the far frontiers of what gravity may one day help us become.

This is not just a book about falling objects. It's about rising minds.

Thank you for letting mine speak to yours.

With deep gratitude and wonder,

—*Pyura Anshuman*

# Prologue

*Before the apple fell, the stars had whispered low,*
*Of threads unseen that bind the high and low.*
*In silence, gravity composes the divine,*
*A song of weight and wonder, space and time.*

**—Pyura Anshuman**

Have you ever found yourself standing beneath the vast, velvety expanse of a night sky, strewn with shimmering stars, and wondered why those distant celestial orbs, the stars, don't simply tumble from their lofty perches, cascading like cosmic waterfalls? Have you ever experienced the marvel of standing amidst nature's symphony, as a solitary leaf, kissed by autumn's gentle breeze, embarks on its poetic descent toward the Earth's welcoming embrace, and in that tranquil, almost ethereal moment, contemplated the invisible and yet omnipresent force that choreographs this delicate ballet? These profound moments of wonder, where the ordinary unfurls into the extraordinary, serve as our portals into the endlessly fascinating and enigmatic realm of gravity—a force so universal, so inherently woven into the very fabric of our reality, that it transcends the mundane and shapes the sublime.

In those distant constellations that have captivated and guided humanity's explorations for millennia, and in the serene descent of a solitary leaf that mirrors our own eventual journey to the welcoming soil, we encounter the telltale imprints of gravity. It is the cosmic adhesive that binds our world and the universe in a perpetual, celestial dance, an unseen maestro orchestrating the intricacies of celestial motions and earthly phenomena alike. Within these seemingly commonplace observations resides a profound inquiry into the unseen forces that silently govern our existence, reminding us that even the most ordinary facets of our world are imbued with the enchantment and mystery of the cosmos.

Welcome, dear readers, to the illuminating odyssey that is "The Gravity Guide: Unveiling the Universe's Hidden Force."

In the hallowed annals of scientific history, there exist iconic moments that forever altered our understanding of the world around us. Picture, if you will, the legendary apple that tumbled from the boughs of an orchard tree, captivating the mind of a young Sir Isaac Newton. In that humble descent, the seeds of a groundbreaking revelation were sown, leading to the formulation of the laws of gravity that govern our physical world. But that was only the beginning.

Fast forward through centuries of intellectual ferment to the early 20th century, where the brilliant mind of Albert Einstein would introduce us to a new dimension of gravity—one that transcended the familiar, gently curving the very fabric of spacetime itself. This radical concept, known as general relativity, thrust us into a cosmic realm where gravity became not just a force but a profound warping of reality, reshaping our understanding of the universe's inner workings.

So, here, within the pages of this guide, we invite you to join us on a voyage through the corridors of history, the realms of physics, and the depths of cosmic enigma. Together, we will unravel the secrets of gravity, explore its far-reaching implications, and gain insight into its role in sculpting the very mosaic of the universe. It is a journey that will challenge your perceptions, ignite your curiosity, and leave you with a profound appreciation for the hidden force that shapes the cosmos we call home.

Embark on this exhilarating voyage with us, as we endeavor to unravel the enigmatic cosmic force that is gravity. In the pages that follow, we will peel away the layers of mystery, revealing the profound and far-reaching influence of this force on the cosmos. From the minuscule, subatomic particles that dance in defiance of classical physics to the majestic grandeur of celestial bodies that adorn the night sky, we will traverse the spectrum of scale and wonder. Together, we will journey through the annals of science, deep into the heart of theoretical physics, and out into the boundless expanse of the universe itself, shedding light on the captivating secrets gravity has held tightly for eons.

Whether you are a curious soul with a penchant for peering into the depths of the universe's mysteries or a seasoned explorer of cosmic enigmas, "The Gravity Guide" beckons you to embark on an enlightening odyssey that delves into the very heart of gravity's profound influence on our existence.

For those of you who have ever cast your gaze upon the night sky and pondered the cosmic forces that govern the celestial dance, this guide promises to be a celestial compass, charting the course through the labyrinthine wonderland of gravity. It is an invitation to expand your understanding, to quench the thirst of your intellectual curiosity, and to unlock the door to a realm where the rules of the ordinary bend and sway in the presence of an extraordinary cosmic force.

And to the seasoned cosmic adventurers, the starry-eyed dreamers, and those who have traversed the intellectual cosmos in pursuit of answers, "The Gravity Guide" extends its hand as a companion on your ongoing journey. It is a resource that seeks to complement your knowledge with new perspectives, deeper insights, and the thrill of discovery. Together, we will navigate through the scientific annals and theoretical landscapes, embarking on an expedition that promises to illuminate the nuances of gravity's influence, leaving no question unanswered and no concept unexplored.

So, whether you are stepping into this cosmic journey with wide-eyed wonder or as a seasoned traveler of the cosmos, "The Gravity Guide" is your ticket to unraveling the profound mysteries of gravity that have shaped our universe since time immemorial. Welcome aboard this voyage of discovery, where every page is an invitation to expand your horizons and deepen your connection with the universe's hidden force.

# Introduction

*Why This Book, Why Now?*

We live each day under the quiet influence of gravity. It is the invisible sculptor of our experiences, keeping our feet grounded, our drinks poured downward, and our planets spinning in silent cosmic harmony. And yet, despite its ever-present role in our lives, gravity remains one of the most misunderstood forces in the universe.

This book was born out of a simple curiosity: **What if we could make gravity not just understandable, but deeply fascinating to anyone—young or old, curious or skeptical?**

From the falling apple that inspired Newton to the mind-bending warps of spacetime revealed by Einstein, gravity has been both a practical puzzle and a profound mystery. In these pages, we take a journey through both realms. We begin with the everyday: why objects fall, how gravity affects your body, and what happens when you jump on the Moon. Then we travel outward into the vast unknown: black holes, time dilation, dark energy, and the search for life beyond Earth.

But this is not just a science textbook. It's a **guide**, designed to spark wonder, connect the cosmic to the personal, and blend rigorous science with hands-on learning. You'll find not only clear explanations and real-world examples but also **engaging experiments** that bring the force of gravity quite literally into your hands.

Whether you're a curious reader, a student, a teacher, or a lifelong learner, this book is your invitation to explore the unseen force that binds the universe—and you—to everything else.

So, turn the page, and let gravity pull you in.

Gravity's everyday manifestations

# PYURA
## BOOKS & RESEARCH

Understanding gravity forges a
path toward better navigation.

# 1

## A Force That Shapes Our World

**"What we know is a drop, what we don't know is an ocean."**
— *Sir Isaac Newton*

Gravity, with its timeless and elegant simplicity, stands as the universal embrace that governs the destiny of every object within the vast and intricate cosmos. It extends its influence over the grandest galaxies, where countless stars twirl in celestial ballets, and it cradles the tiniest particles, where the very essence of existence takes shape. As we embark on this intellectual voyage, both as scientists and philosophers of the cosmos, I extend to you an invitation to delve deep into the contemplative realm and fathom the profound impact of this celestial force on the intricate mosaic of our existence.

Imagine, if you will, the grandeur of galaxies, each one a sprawling city of stars, held in their cosmic dance by the gentle yet unyielding grasp of gravity. This force is the cosmic conductor of this grand symphony, orchestrating the movements of celestial bodies on a canvas as vast as the universe itself. Yet, gravity's reach extends far beyond the bounds of galaxies, down to the subatomic level where particles engage in their quantum waltz, subject to the same universal force that guides the stars.

And so, as we embark on this exploration, we uncover that gravity is not merely an invisible thread pulling us inexorably downward, but a profound and fundamental element that sculpts the very contours of our reality. It is the architect of worlds, the sculptor of spacetime, and the silent maestro of the cosmic orchestra. Within the simplicity of its embrace lies a universe of complexity waiting to be unveiled, a universe that beckons us to ponder its mysteries and marvel at the celestial symphony that unfolds before our eyes.

Consider, if you will, the magnificent mosaic that is our planet Earth, where the invisible hand of gravity plays the role of the master sculptor. It gently molds and shapes the contours of our world, giving rise to majestic mountains that pierce the sky, carving deep valleys that cradle winding rivers, and forming vast oceans that stretch to the horizon. Gravity, in its patient and persistent manner, is the artist behind these geological wonders, etching its masterpiece upon the canvas of our planet's surface.

Beyond our terrestrial abode, gravity extends its influence to the far reaches of the cosmos, where celestial bodies engage in cosmic ballets that defy time itself. It is the cosmic choreographer, orchestrating the graceful movements of planets around their suns, stars within their galaxies, and galaxies within the vast expanse of the universe. These cosmic dances, guided by the invisible strings of gravity, unfold across epochs, painting a picture of celestial harmony that we can only begin to comprehend.

And yet, despite the profound impact it has on our world and the cosmos at large, gravity often remains in the background, an unassuming force that goes unnoticed in our daily lives. It is the unbreakable bond that ties us to our home in the cosmos, the force that keeps our feet firmly planted on the ground, and the reason the moon traces its elegant path across the night sky. In its subtle and ever-present embrace, we find a force so intricately woven into the fabric of reality that we may take its wonders for granted. But as we embark on this cosmic journey, we are invited to pause and marvel at the intricate dance of gravity, a force that sculpts worlds and guides the cosmos, shaping the very essence of our existence.

However, let us not constrain gravity to the role of a mere force of attraction; it is a cosmic storyteller of unparalleled eloquence. Within its silent embrace, it pens the epic saga of the universe, a narrative that spans from the cosmic cradle of star birth, where the relentless clutches of gravity coax nebulous clouds into fiery life, to the mesmerizing and enigmatic realms of black holes, where even the fastest travelers, beams of light, succumb to its unfathomable grasp.

In the celestial drama that unfolds, gravity is the principal character, orchestrating the movements of planets, stars, and galaxies across the vast cosmic stage. It writes the chronicles of cosmic evolution, shaping the destinies of celestial bodies and sculpting the very landscapes of the cosmos. Each celestial dance, from the intricate pirouettes of binary stars to the spiraling waltz of galaxies, is choreographed by gravity's invisible hand, leaving behind a rich narrative that astronomers and physicists strive to decipher.

But gravity's storytelling prowess extends beyond the celestial realm. It is also the cosmic quill that bends the very fabric of spacetime itself. Through the lens of Einstein's theory of general relativity, we uncover its most profound chapters—a tale of time dilation, where clocks tick at different rates depending on the strength of gravity's pull, and a narrative of spacetime warping, where massive objects like stars and black holes create ripples that traverse the cosmos as gravitational waves, carrying with them the echoes of cataclysmic events.

# Massive objects cause the warping of spacetime, distorting the grid of reality.

So, as we embark on this cosmic journey, let us embrace gravity not only as the force that binds us to our world but as the masterful storyteller of the cosmos. Within its narrative lies the key to unlocking the secrets of the universe, revealing the intricate interplay of forces and phenomena that shape the grand spectacle of existence.

As our odyssey into the depths of gravity continues, we stand at a remarkable crossroads where science converges with philosophy, and questions about the very fabric of reality beckon our contemplation. It is within this juncture that we grapple with profound inquiries about the essence of space, the elusive nature of time, and the intricate web of interconnectedness that binds all things in the cosmos. Gravity, as our guiding star on this intellectual journey, propels us into an exploration that transcends the boundaries of disciplines, inviting us to ponder not only the mechanics of the universe but also the profound metaphysical questions that lie at its core.

In the silent embrace of gravity, we are confronted with the eternal mystery of existence itself. It is a force that extends its influence across the vast cosmic expanse, shaping the destinies of galaxies, stars, and planets. Yet, it is also a force that tethers us to our terrestrial home, reminding us of our humble place in the grand cosmic mosaic. As we navigate the labyrinthine corridors of gravity's influence, we are compelled to reflect upon our role in the cosmic drama.

Gravity, often referred to as the invisible hand of the universe, emerges not only as the sculptor of our world but also as the masterful composer of our cosmic symphony. It orchestrates the celestial movements that grace the night sky, dictating the rhythms of the cosmos with a precision that leaves us in awe. In its subtlety and grandeur, it invites us to contemplate the harmonious interplay of forces and elements that give rise to the intricate dance of existence.

Thus, as we venture deeper into this cosmic exploration, we find that gravity is not merely a scientific concept but a profound window into the very heart of the cosmos. It is a force that bridges the realms of science and philosophy, inviting us to ponder the mysteries of space and time, the unity of all things, and our place in the vast cosmic narrative. In the embrace of gravity, we discover a cosmic revelation that reshapes our understanding of the universe and prompts us to contemplate our role as participants in the magnificent dance of existence.

## 1.1 Everyday experiences related to gravity

Gravity, often described as the silent architect of our existence, manifests itself in a multitude of ways, subtly yet profoundly influencing the mosaic of our daily lives. It is a force so woven into the very fabric of our reality that we seldom pause to contemplate its omnipresent effects. From the simplest acts to the grandest phenomena, gravity is an ever-watchful companion that leaves its indelible mark.

Consider the gentle descent of a leaf, spiraling earthward from the heights of a tree. This elegantly choreographed descent is a testament to gravity's embrace, its invisible hand guiding the leaf's journey as it flutters to the ground. Or, when we step out of bed each morning, our feet meet the floor with unwavering predictability, courtesy of gravity's constant pull. It's the reason why our morning rituals are marked by a sense of groundedness, as we navigate our homes and communities under the watchful gaze of this unseen cosmic force.

Yet, gravity's influence extends far beyond the confines of our homes. It shapes the landscapes we traverse, carving out majestic mountain ranges and sculpting valleys with an unerring touch. It governs the graceful orbits of celestial bodies, orchestrating cosmic dances that span millennia. In essence, gravity is not a passive bystander but an active participant in the unfolding narrative of our world.

From the tiniest raindrop that falls from the sky to the awe-inspiring majesty of a soaring eagle, gravity is the unifying thread that binds all things to the Earth. It is the force that keeps our oceans in check, ensures the stability of our skyscrapers, and allows us to savor the simple pleasure of a refreshing sip of water. Gravity is the ever-present companion that shapes our experiences, influences our understanding of the physical world, and, in its silent and unassuming manner, reminds us of our profound connection to the cosmos.

Consider the simple act of dropping an object. It's a seemingly mundane occurrence, yet it provides us with a tangible connection to the force that governs the cosmos. When we release an item, whether it's a pen, an apple, or a feather, we witness gravity's inexorable pull as it guides the object towards the Earth's surface. This everyday encounter serves as a gentle reminder that, in the grand mosaic of the universe, we are all subject to the same force.

The act of dropping objects, whether intentional or accidental, is a universal phenomenon that offers a tangible illustration of gravity's influence. When we release an object from our grasp, it begins its descent towards the Earth's surface. This seemingly straightforward occurrence is governed by the fundamental principle of gravity, articulated by Sir Isaac Newton over three centuries ago.

In this act, we observe the universal law of gravitation in action. Every object with mass attracts every other object with mass, and the strength of this attraction is directly proportional to the mass of the objects and inversely proportional to the square of the distance between them. As a result, the Earth, with its immense mass, exerts a gravitational pull on all objects within its vicinity. When we release an object, it accelerates towards the Earth under the influence of gravity, obeying this law with remarkable precision.

Dropping an object illustrates
gravity's influence.

This everyday experience not only connects us to the cosmos but also serves as a foundation for our understanding of the physical world. It's the reason why we feel objects fall when we let go of them and why celestial bodies, from apples to planets, follow predictable paths in their orbits around the Sun. These seemingly mundane moments, like dropping a ball or watching a leaf float to the ground, are profound reminders that gravity is an ever-present force that governs the motion of objects on Earth and throughout the universe.

Walking, too, offers an intimate interaction with gravity. With each step we take, our bodies remain anchored to the Earth by this invisible tether. When we take a step, we exert a force against the ground, and the ground exerts an equal and opposite force, propelling us forward. This interaction between our bodies and the Earth's surface is a result of the gravitational pull of our planet. It's a sensation we often take for granted, but it underscores the ever-present influence of gravity in our lives. As we stroll along, we can appreciate how this force keeps us grounded, enabling us to explore our world and navigate the terrain beneath our feet. These seemingly ordinary experiences serve as a starting point for our exploration of gravity's profound implications, as we venture deeper into the scientific and philosophical realms of this omnipresent cosmic force.

In walking, we become aware of the resistance offered by gravity, which we must overcome with each step. This resistance gives us stability, allowing us to maintain an upright posture and explore our environment. Moreover, it is gravity that enables us to have a sense of "down" and "up," making it possible for us to orient ourselves and navigate our surroundings effectively.

Gravity, acting as an invisible anchor, keeps us connected to the Earth. It ensures that we remain firmly grounded and that we can traverse the terrain beneath us with relative ease. Without this gravitational force, our world would be vastly different, and simple actions like walking would be impossible.

Beyond the physical aspects, walking also has psychological and philosophical implications related to gravity. It reflects our connection to the Earth and highlights the notion that, in our daily lives, we are inextricably linked to the forces that govern the cosmos. The act of walking, which may seem routine, serves as a profound reminder of our place in the grand cosmic scheme, where gravity is the ever-present force that tethers us to our home planet and the broader universe.

## 1.2 Understanding gravity can answer common questions and improve daily life

Understanding gravity is like holding a key to unlock a treasure trove of knowledge and practical insights that can significantly enhance our daily lives and provide answers to common questions. Here's how:

**Weight Management:** Gravity, that silent force which we often take for granted, is the very reason we experience weight on Earth's surface. It is a fundamental concept that, when understood, can empower us to make informed decisions about our health and overall well-being. In essence, gravity is not just a cosmic phenomenon but an integral part of our daily lives, and comprehending its role unveils a valuable key to maintaining our physical fitness and vitality.

At the heart of this understanding lies the realization that gravity is a constant presence, ceaselessly pulling us towards the Earth's core. This force creates the sensation of weight we experience when we step on a scale or lift an object. With this knowledge in hand, we gain a deeper appreciation for the importance of activities like exercise. Regular physical activity becomes a counterbalance to gravity's relentless pull, helping us maintain a healthy weight and preserve our overall fitness.

Moreover, understanding gravity's role in our lives provides us with a compass for making wise choices about our health. We become aware that a sedentary lifestyle, where we defy the laws of gravity by remaining motionless for extended periods, can lead to a host of health issues. These can range from weight gain and muscle atrophy to reduced cardiovascular health. Armed with the knowledge that gravity is an ever-present force that shapes our physical reality, we are inspired to make choices that align with our well-being, prioritizing activities that keep us in harmony with this cosmic force.

Gravity's pull gives rise to the sensation of weight.

In essence, gravity becomes a partner in our journey toward better health, guiding us to make choices that honor the intricate dance between our bodies and the force that governs our world. By comprehending this fundamental concept, we gain not only a deeper understanding of our physical selves but also a powerful tool for managing our fitness and overall health, allowing us to navigate the challenges of gravity with grace and vitality.

**Navigation:** Gravity, in its subtle yet all-encompassing presence, plays a pivotal role in shaping our sense of direction and orientation in the world. This profound influence extends far beyond mere physics; it permeates our daily lives, impacting everything from our ability to read maps to our reliance on GPS technology. As we dive deeper into this gravitational understanding, we unearth a wealth of knowledge that proves invaluable, particularly for those intrepid souls who embark on journeys, traverse unfamiliar terrain, or navigate the intricate paths of our globe.

Imagine for a moment the act of reading a map, an essential skill for travelers and adventurers alike. With the knowledge of gravity's relentless pull toward the Earth's center, we gain a newfound appreciation for the accuracy and reliability of such maps. We understand that every contour, riverbed, and mountain range etched on that piece of paper is a testament to the way gravity has sculpted the Earth's surface over eons. This understanding transforms the map from a mere collection of lines and symbols into a vivid representation of the forces that have shaped our planet. It allows us to interpret it with a depth and clarity that transcends the flatness of the page.

Moreover, gravity's role extends to our use of compasses, those trusty navigational tools that have guided explorers for centuries. When we grasp the fundamental connection between gravity and our orientation on Earth, we gain insights into how compasses function. These devices, which rely on the magnetic field generated within our planet, work in tandem with gravity to provide us with accurate directional information. Understanding this synergy between magnetic forces and gravity enhances our ability to navigate with precision, whether we're deep in the wilderness, sailing on the open sea, or simply finding our way through a bustling city.

In today's technologically advanced world, the influence of gravity on navigation is further magnified through GPS devices. These remarkable tools, which pinpoint our exact location on Earth, operate on principles grounded in the interplay of satellites, time, and the influence of gravity. The satellites that make up the GPS constellation follow carefully calculated orbits, with their internal atomic clocks ticking in perfect harmony. However, due to the time dilation effects of gravity (a phenomenon predicted by Einstein's theory of general relativity), these clocks experience slight differences in their rates. Without precise corrections for these gravitational effects, the accuracy of GPS systems would be compromised. Understanding gravity's role in time dilation becomes crucial for the accuracy of GPS navigation, ensuring that we reach our destinations with precision.

# Gravity's role in GPS accuracy ensures precise navigation.

Therefore, comprehending the intricate relationship between gravity and our sense of direction and orientation not only enriches our understanding of the natural world but also equips us with the tools needed for accurate navigation. Whether we're explorers in search of distant lands, hikers navigating challenging terrain, or urban travelers finding our way through bustling city streets, this knowledge becomes an indispensable compass guiding us on our journeys. Gravity, the invisible force that pulls us toward the Earth, becomes the silent collaborator in our adventures, offering us the knowledge and insight to explore our world with confidence and purpose.

**Engineering and Architecture:** Gravity, often regarded as nature's unwavering force, serves as the unseen but ever-present collaborator in the field of architecture and engineering. It is the silent partner that architects and engineers rely upon to design structures of immense complexity and elegance, from towering skyscrapers that scrape the heavens to bridges that span great chasms. Within this gravitational understanding lies the blueprint for crafting safe, resilient, and structurally sound edifices that weave themselves into the fabric of our urban landscapes and daily lives.

The influence of gravity on architectural and engineering endeavors is undeniable. When architects conceive their designs, and engineers draft their plans, they do so with an acute awareness of gravity's unyielding grasp. Every choice, from the materials used to the shapes and forms of structures, is made with the overarching need to counteract this force. For architects, this means crafting buildings with well-balanced proportions that can withstand the relentless downward pull of gravity. For engineers, it entails designing support systems and foundations that can bear the immense loads imposed by the structures they underpin.

Consider, for instance, a towering skyscraper reaching towards the heavens. Each floor, each beam, and each column must be meticulously designed and constructed to counteract the force of gravity. Engineers calculate the precise distribution of weight and stress, ensuring that the building's structural elements can endure the continuous pull towards the Earth's center. The elegant dance between form and function, aesthetics and engineering, hinges on their intimate understanding of gravity's role in shaping the built environment.

This gravitational knowledge becomes especially critical when constructing infrastructure that supports not only the weight of the structure itself but also the weight of the lives and activities it shelters. Bridges that span vast rivers, tunnels that burrow through mountains, and highways that crisscross the landscape all rely on an in-depth comprehension of gravity's effects on materials and structures. Engineers must account for factors such as compression, tension, and torsion to ensure that these vital conduits remain safe and reliable, regardless of the forces they encounter.

# Gravity plays a crucial role in engineering and architecture.

The collaboration between architects, engineers, and gravity is a symphony of science and creativity, precision and artistry. It is a testament to human ingenuity and the quest to harmonize our built environment with the natural forces that govern our world. This gravitational understanding is the cornerstone of the structures we depend on daily, from the home where we dwell to the bridges we traverse. It ensures not only their stability and durability but also the safety and well-being of the communities they serve. In the realm of architecture and engineering, gravity becomes both the challenge and the inspiration, shaping our urban landscapes and enriching our lives.

**Astronomy and Space Exploration:** Gravity, the universal maestro, extends its far-reaching influence far beyond the boundaries of our blue planet, conducting the grand cosmic orchestra that shapes the movements of celestial bodies in the vast expanse of the cosmos. This profound understanding of gravity has not only expanded our comprehension of our own solar system but has also provided answers to some of the most profound questions about the mechanisms that govern the heavens.

One of the celestial puzzles that gravity has helped us solve is the intricate dance of the planets in our solar system. It was Sir Isaac Newton who, with his revolutionary law of universal gravitation, illuminated the precise mathematical relationships that dictate the elliptical orbits of planets around the Sun. This revelation not only confirmed our heliocentric model of the solar system but also unlocked the means to predict the positions of planets with extraordinary accuracy. The ability to forecast planetary positions and predict celestial events like eclipses became invaluable for astronomers and navigators alike, ushering in a new era of celestial understanding and exploration.

Furthermore, our comprehension of gravity has cast light on the phases of the Moon, unraveling the lunar cycles that have mystified humanity for eons. The gravitational interplay between Earth, the Moon, and the Sun results in the mesmerizing dance of lunar phases, from the waxing crescent to the full moon and back to the new moon. Understanding this celestial choreography has not only enriched our cultural heritage but also served as a fundamental tool for calendars and timekeeping throughout history.

Our understanding of gravity guides space exploration throughout the solar system.

In the context of space exploration, gravity emerges as the keystone to planning missions that venture beyond our home planet. When sending spacecraft to other celestial bodies, such as the Moon, Mars, or distant asteroids, we must meticulously calculate the gravitational forces at play. These calculations determine the spacecraft's trajectory, its entry into orbit around the target body, and even the precision of its landing. Gravity's guiding hand ensures that these missions can navigate the complexities of space and fulfill their scientific objectives.

Furthermore, the profound influence of gravity extends its reach to the furthest corners of the universe. The gravitational pull between galaxies shapes the large-scale structure of the cosmos and influences the motion of stars within them. Through the study of gravity, astrophysicists have uncovered the hidden mass of dark matter, an enigmatic cosmic ingredient that comprises a significant portion of the universe's substance. Additionally, the discovery of dark energy, another mysterious cosmic force responsible for the accelerated expansion of the universe, was made possible by gravitational observations.

Our understanding of gravity has transcended the boundaries of our home planet, becoming a cosmic compass that guides us through the intricacies of the solar system and beyond. It has enabled us to decode the celestial mysteries of planetary orbits, lunar phases, and the vast cosmic web of galaxies. As we continue to explore the universe, gravity remains our steadfast companion, illuminating the path to the stars and revealing the profound interconnectedness of all things in the cosmos.

**Timekeeping:** Einstein's theory of relativity unveiled a profound and mind-bending connection between gravity and time. This revelation, while astonishing from a theoretical perspective, carries practical implications that touch our daily lives in unexpected ways, particularly in the realm of technology.

One of the most tangible manifestations of this interplay between gravity and time is in the functioning of the Global Positioning System (GPS). GPS systems rely on a constellation of satellites orbiting the Earth, each equipped with precise atomic clocks. These clocks, however, are subject to the effects of both special relativity and general relativity, which dictate that time passes differently in regions of varying gravitational strength.

Specifically, according to general relativity, time flows more slowly in stronger gravitational fields. Since the Earth's gravitational field is stronger at its surface than in the space where the satellites orbit, the atomic clocks aboard the GPS satellites experience time dilation—they tick slightly faster relative to clocks on the Earth's surface. If this relativistic effect were not taken into account, the timekeeping of the satellites and, consequently, the accuracy of GPS positioning would drift apart over time.

Understanding the nuances of gravity's influence on time is the key to mitigating these relativistic effects and maintaining the precision of GPS systems. The satellites are programmed with the necessary corrections to ensure that their signals accurately reflect the time on Earth's surface. This ongoing synchronization is crucial for the pinpoint accuracy of

GPS-based navigation and location services, which we rely upon for everything from finding our way in unfamiliar cities to tracking valuable shipments around the globe.

# Understanding gravity's effects on time is essential for the precise operation of modern timekeeping systems.

Moreover, this marriage of gravity and time dilation has practical applications beyond GPS technology. It underscores the importance of precise timekeeping in various fields, from telecommunications to financial transactions. Accurate time synchronization ensures the smooth operation of our interconnected global systems, bolstering the efficiency and reliability of technologies that permeate our modern world.

The understanding of gravity's impact on time, as illuminated by Einstein's theory of relativity, has far-reaching practical consequences in our technologically driven society. It enables us to harness the insights of modern physics to fine-tune the instruments that define our daily lives, ensuring that the intricate interplay of gravity and time remains a reliable guide in our quest for precision and accuracy.

**Innovation and Technology:**

The profound insights garnered from our understanding of gravity have had a transformative impact on a wide range of technologies, sparking innovation and improving the quality of life across diverse industries. Gravity, often perceived as a force that shapes the cosmos, is harnessed ingeniously to power technologies that span from healthcare to scientific research and beyond. Here, we delve into the remarkable ways in which gravity-based technologies have influenced our world:

*Drug Delivery Systems:* In the field of healthcare, gravity plays a pivotal role in the development of advanced drug delivery systems. Intravenous (IV) therapy, for instance, relies on gravity to regulate the flow of medications and fluids directly into a patient's bloodstream. Gravity-driven infusion devices provide a controlled and reliable means of administering treatments, ensuring that patients receive precise dosages in a safe and efficient manner. These gravity-based systems have become a cornerstone of modern healthcare, from hospitals to home-based care, enhancing the management of various medical conditions.

*Sensors and Instruments:* Gravity-based technologies underpin an array of sensors and instruments used in scientific research. Gravitational sensors, such as gravimeters, exploit the Earth's gravitational field to measure minute changes in gravity, enabling precise geophysical studies. Furthermore, gravitational wave detectors, like the Laser Interferometer Gravitational-Wave Observatory (LIGO), employ incredibly sensitive instruments to detect ripples in spacetime caused by cataclysmic events, such as the collision of black holes. These groundbreaking technologies have opened a new era in astrophysics, offering direct observations of phenomena previously hidden from human view.

*Space Exploration:* The exploration of space heavily relies on our understanding of gravity. Spacecraft and satellites are equipped with propulsion systems that exploit gravitational forces to navigate, enter orbit, and reach distant destinations. Furthermore, space missions often utilize gravitational assists from celestial bodies to conserve fuel and gain momentum.

These ingenious maneuvers, such as gravitational slingshots around planets, have enabled ambitious missions like the Voyager probes and the New Horizons spacecraft to explore the farthest reaches of our solar system.

**Advanced Manufacturing:** Gravity-based technologies are employed in advanced manufacturing processes. Techniques such as gravure printing, which relies on the gravitational flow of ink onto a printing substrate, are used in industries ranging from packaging to electronics production. Moreover, precision gravity casting, a method that leverages gravity to fill molds with molten materials, is used to create intricate and high-quality metal components for various applications, including aerospace and automotive industries.

**Renewable Energy:** The concept of gravity is integral to various renewable energy technologies. Hydropower, for instance, harnesses the gravitational pull of water as it flows downhill to generate electricity. Gravity-based energy storage systems, like pumped storage hydropower, store excess energy during periods of low demand and release it when needed, contributing to grid stability in renewable energy networks.

Our understanding of gravity has led to a remarkable array of gravity-based technologies that have revolutionized multiple industries and facets of our lives. From healthcare to scientific research and beyond, gravity remains a silent yet essential partner in driving innovation and progress. These technologies, born from the profound insights into the nature of gravity, continue to push the boundaries of what is possible, enhancing our quality of life and deepening our understanding of the universe.

**Basic Curiosity:** At its core, our comprehension of gravity serves as a philosophical journey into the mysteries of the universe, satisfying our innate curiosity about the fundamental forces that shape the cosmos and our place within it. This exploration goes beyond measurements; it delves into the very essence of our existence and sparks profound questions that have captivated human minds for generations.

One of the most fundamental queries that understanding gravity addresses is the simple yet profound "Why do objects fall?" This age-old question, posed by thinkers from ancient civilizations to the Renaissance era, has been central to our understanding of the physical world. Sir Isaac Newton's groundbreaking work in the 17th century, culminating in the formulation of the law of universal gravitation, provided an unprecedented answer. Gravity, he revealed, is the invisible force that pulls objects toward one another—the cosmic embrace that governs the destiny of every particle in the universe. This revelation not only demystified the act of falling but also illuminated the unifying principle that connects the celestial and terrestrial realms.

In a similar vein, understanding gravity addresses the perennial question of "What keeps celestial bodies in motion?" The ancient Greeks pondered the intricate dance of the heavens, as stars, planets, and moons moved with apparent precision across the night sky. The emergence of Newton's laws of motion and universal gravitation provided the framework to decipher this celestial choreography. It revealed that gravity is the force that binds celestial bodies in an eternal embrace, shaping the orbits of planets around stars and moons around planets. This profound insight into the workings of the universe not only enriched our understanding of the cosmos but also kindled the flames of scientific exploration.

These inquiries into the nature of gravity, while rooted in the physical sciences, transcend mere scientific curiosity. They touch upon the very essence of our existence and our innate drive to explore the unknown. The pursuit of answers to questions like "Why do objects fall?" and "What keeps celestial bodies in motion?" is not confined to laboratories and observatories; it reverberates through the corridors of philosophy, inspiring contemplation about the nature of reality and our place within it.

As we unravel the mysteries of gravity, we embark on a journey that unites the realms of science and philosophy, sparking a sense of wonder and curiosity that transcends generations. It is a journey that propels us to look up at the night sky with awe and to ponder our place in the grand mosaic of the cosmos. In understanding gravity, we uncover not just the secrets of the universe but also the enduring human spirit of inquiry, discovery, and exploration. It is a force that, in more ways than one, pulls us toward a deeper understanding of the world and the universe beyond.

# Gravity is the force that keeps celestial bodies in motion.

In sum, gravity serves as a foundational principle that underpins numerous aspects of our daily existence. By grasping its intricacies, we gain valuable insights into the physical world, address practical questions, and open doors to technological advancements that shape our lives for the better. Gravity is not just a force of nature but a guiding light that illuminates our path toward a deeper comprehension of the universe and the improvement of our daily lives.

# 2

# The Surprising Story of Falling Objects

"The important thing is not to stop questioning.

Curiosity has its own reason for existing."

— *Albert Einstein*

Imagine a world where objects don't fall, where feathers drift upwards, and apples remain suspended mid-air. It's a world that defies our everyday reality, and yet, understanding the profound story of falling objects brings us closer to unraveling the mysteries of gravity, a force that shapes our universe. In this captivating narrative, we embark on a journey through history, science, and the boundless cosmos to explore the surprising and enchanting tale of objects in free fall.

At the heart of this story is Sir Isaac Newton's legendary encounter with a falling apple, which sparked a scientific revolution and led to the formulation of the universal law of gravitation. We delve into the annals of history to witness pivotal moments when brilliant minds grappled with the enigma of falling objects, from Galileo's daring experiments atop the Leaning Tower of Pisa to the moonlit musings of Johannes Kepler. Their groundbreaking insights laid the foundation for our modern understanding of gravity and set the stage for the exploration of the cosmos.

But the story doesn't end on Earth's surface. We journey beyond our planet to explore the astonishing effects of gravity on celestial bodies. From the graceful orbits of planets around the Sun to the cataclysmic collisions of galaxies in the depths of space, we discover how gravity shapes the grand cosmic ballet of the universe.

As we unravel the surprising story of falling objects, we encounter not only the profound influence of gravity but also the timeless allure of curiosity and discovery. It's a story that bridges the gap between the everyday and the extraordinary, inviting us to ponder the forces that govern our world and the wonders that await our exploration. Join us on this extraordinary voyage as we uncover the secrets of falling objects, and in doing so, gain a deeper appreciation for the captivating force that keeps us grounded while allowing the cosmos to dance in the boundless expanse of the universe.

## 2.1 The history of gravity's discovery

The story of gravity's discovery is a journey filled with intrigue, experimentation, and moments of profound insight. While it involves complex scientific concepts, it's a tale that can be both accessible and engaging.

Our story begins in ancient times when philosophers and scholars pondered the mysteries of falling objects. The great philosopher Aristotle, for instance, proposed that heavy objects fell faster than lighter ones, a notion that held sway for centuries. However, it was a brilliant thinker of the Renaissance, Galileo Galilei, who ushered in a new era of understanding.

In the late 16th century, Galileo stood atop the Leaning Tower of Pisa and, according to legend, dropped two different-weight objects. To the surprise of many, both hit the ground simultaneously, challenging Aristotle's beliefs. This simple yet groundbreaking experiment marked the birth of modern physics and our journey toward comprehending gravity's true nature.

Fast forward to the late 17th century, where we encounter the remarkable Sir Isaac Newton. Observing an apple falling from a tree in his orchard, Newton had a moment of profound insight. He realized that the same force responsible for the apple's fall also governed the motion of celestial bodies. This insight led to his formulation of the law of universal gravitation, a groundbreaking concept that described how every object with mass attracts every other object. It was this revelation that forever changed our understanding of gravity.

# The history of gravity's discovery

As our story progresses, we witness the ongoing refinement of our understanding of gravity, with Albert Einstein's theory of general relativity as a key milestone. This revolutionary theory described gravity not as a force but as the curvature of spacetime by massive objects. It predicted phenomena like the bending of starlight around the Sun, which was famously confirmed during a solar eclipse in 1919, catapulting Einstein to global fame.

Today, the story of gravity's discovery continues with ongoing research and exploration. Scientists use cutting-edge technology to study gravitational waves, elusive ripples in spacetime caused by cataclysmic cosmic events, furthering our comprehension of this fascinating force.

Throughout this journey, the discovery of gravity transcends science alone. It's a story of human curiosity, perseverance, and the relentless quest for knowledge. It reminds us that even the simplest observations, like a falling apple, can lead to profound revelations about the fundamental forces that shape our universe. In this tale, accessible and engaging, we find inspiration to continue exploring the mysteries of our cosmos and to embrace the wonder of discovery.

## 2.2 Stories of early scientists and their discoveries

Let's take a closer look at some early scientists and their gravity-related discoveries, bringing their stories to life in a relatable way:

**Galileo Galilei and the Leaning Tower of Pisa:**

Let's step back in time to the late 16th century, an era of profound curiosity and intellectual exploration. Picture yourself atop a slightly leaning tower, the wind rustling your clothing as you clutch two objects of different weights in your hands. You are not alone in this endeavor; you are in the company of the visionary scientist and thinker, Galileo Galilei. While you don't have a modern smartphone to document the moment, what unfolds here will become an iconic episode in the annals of scientific history.

With unwavering determination, Galileo conducts a daring and audacious experiment. In one hand, he holds a heavy object, while in the other, a light one. Both objects are poised at the tower's edge, ready to meet their destiny. As you watch with bated breath, Galileo releases both objects simultaneously, and they plummet earthward. Time seems to slow as they descend, and to everyone's astonishment, the heavy and light objects hit the ground at the exact same time.

This seemingly simple yet brilliantly conceived experiment defied the prevailing belief of the time that heavier objects fell faster—a belief rooted in ancient philosophy and perpetuated for centuries. Galileo's groundbreaking experiment became an "Aha!" moment that resonated through the corridors of science. It was a transformative instant that shattered the conventional wisdom of the era and sent ripples of enlightenment throughout the world.

Galileo's bold act marked the beginning of a scientific revolution—one that would eventually lead to the formulation of the laws of motion and the modern understanding of gravity. His tower experiment challenged orthodoxy, ushering in an age of empirical inquiry and a shift toward a more evidence-based approach to understanding the natural world. It laid the foundation for the profound insights of Sir Isaac Newton and Albert Einstein, who would further unravel the enigma of gravity.

Galileo dropping spheres of different weights from the Leaning Tower of Pisa to demonstrate that they fall at the same rate.

This moment atop the leaning tower was not just a leap forward in the realm of physics; it was a testament to the power of human curiosity and the unyielding spirit of scientific inquiry. It showed that even in the face of conventional wisdom, a single experiment could change the course of history and transform the way we perceive the universe. Galileo's courageous experiment serves as an enduring reminder that in the pursuit of knowledge, there are no limits to what we can achieve when we dare to question and explore the mysteries of the cosmos.

**Isaac Newton and the Falling Apple:**

Let's transport ourselves to a quaint English orchard, where nature's serenity envelops the landscape. Among the gnarled branches laden with ripe apples stands a young Isaac Newton, an inquisitive mind with a penchant for observing the world around him. He is on the cusp of an intellectual revelation that will forever alter our understanding of the cosmos.

As Newton gazes upward, his attention is drawn to an apple perched high in the tree's branches, bathed in the golden glow of the sun. With each passing moment, the apple's descent becomes inevitable, guided by an unseen hand—a hand that will soon redefine the very laws of nature.

The apple detaches from its arboreal perch, beginning its graceful descent toward the Earth. Newton's keen eyes track its every move, and as the apple nears the ground, he ponders a question that would echo through the ages: "Why does the apple fall in such a precise manner?" In that contemplative moment, the seeds of an idea take root—a notion that the force responsible for this apple's descent might also be at play on a grander scale, governing the motion of celestial bodies.

# Isaac Newton discerning the law of gravitation from the falling of an apple.

This unassuming apple, gently cradled by Newton's orchard, serves as the catalyst for one of the most remarkable revelations in the history of science. It sparks Newton's curiosity about the Moon's orbit around the Earth and, in time, leads to the articulation of the universal law of gravitation. Newton's epiphany is the quintessential example of how a seemingly ordinary event can trigger extraordinary insights. It's a testament to the power of keen observation, contemplation, and the boundless potential of the human intellect to unravel the mysteries of the universe.

The iconic image of Newton beneath the apple tree, often depicted with a gravity-defying apple, symbolizes not just a moment of inspiration but a profound shift in our comprehension of the cosmos. This singular event in an orchard in Woolsthorpe, England, reverberates through the annals of science, reminding us that the simplest observations can lead to the most profound discoveries. It is a reminder that the universe's most enigmatic secrets may be concealed in the everyday phenomena that surround us, waiting to be uncovered by curious minds.

**Albert Einstein and the Bending Light:**

Imagine journeying through the corridors of time to the early 20th century, a time when the world was on the brink of a scientific revolution, and the name Albert Einstein was about to become synonymous with intellectual brilliance. Einstein, with his unruly hair and penchant for profound ideas, stands at the forefront of this intellectual transformation, poised to unveil a theory that will reshape our understanding of the very fabric of the universe.

One of Einstein's most captivating ideas revolves around gravity, and it's a concept that pushes the boundaries of our imagination. Picture a massive celestial body, like our Sun, suspended in the cosmic mosaic of space. Einstein postulates that this colossal entity possesses such gravitational might that it can warp the very essence of reality—light itself. This audacious proposal suggests that when light passes near a massive object, its path is bent, much like a river meandering through a canyon. It's a mesmerizing notion that challenges conventional wisdom and invites us to perceive the universe through a new lens.

The moment of truth arrives during a solar eclipse in 1919, an event that captures the collective imagination of scientists and stargazers alike. Armed with telescopes and unwavering determination, scientists across the globe train their instruments on the darkened sky. As the moon veils the Sun's brilliance, a celestial canvas is unveiled, punctuated by distant stars. It is in this cosmic theater that the stage is set for a historic experiment—a test of Einstein's audacious theory.

The observers scrutinize the positions of these distant stars, and to their astonishment, they witness a celestial ballet. The stars, which should have occupied fixed positions in the sky, appear to shift ever so slightly. It is a phenomenon that can only be explained by the gravitational embrace of the Sun, precisely as Einstein had predicted. This groundbreaking moment catapults Einstein to international acclaim, and he ascends to celebrity status. More importantly, it confirms his theory of general relativity—a theory that upends classical physics and inaugurates a new era of understanding the cosmos.

Einstein theorizes that a massive star can bend the very fabric of space and time, causing light to curve around it.

Einstein's journey through the annals of history and his unruly hair become emblematic of a transformative epoch in science. His audacious ideas, like the bending of light by gravity, remind us that the universe is full of surprises and that even the most fundamental aspects of nature can be redefined by the boundless imagination of a singular genius. The solar eclipse of 1919 stands as a testament to the power of human curiosity, observation, and the audacity to challenge the status quo. It is a reminder that science is a journey of perpetual discovery, where each new revelation invites us to glimpse the universe with fresh eyes and a sense of wonder.

These stories remind us that even the greatest scientific breakthroughs often begin with everyday observations and questions. Galileo, Newton, and Einstein were not only brilliant scientists but also curious individuals who dared to challenge conventional wisdom and explore the mysteries of the world around them. Their journeys inspire us to embrace curiosity, observe the world with wonder, and seek answers to the questions that spark our imagination. In the end, it's these moments of curiosity and discovery that drive scientific progress and shape our understanding of gravity and the universe.

# 3

# The Gravity in Your World

**"Gravity is not just a force.**

**It is the very structure of spacetime itself."**

— *Stephen Hawking*

Gravity, the universal force that shapes the cosmos, is not a distant concept confined to the realms of space and physics. It's a force that intimately weaves itself into the very fabric of our everyday lives, affecting the world around us in ways both subtle and profound.

Consider your morning ritual. As you step out of bed, your feet meet the floor with a reassuring sense of familiarity. This is gravity at work, anchoring you to the Earth, providing you with stability and a reference point in your daily life.

When you pour a cup of coffee, gravity gently guides the liquid from the pot into your cup. It ensures that the coffee flows downwards, filling your cup rather than floating aimlessly in the air.

Even the act of enjoying a meal is a testament to gravity's influence. Whether you're savoring a hearty stew or munching on a crisp salad, gravity ensures that your food remains on your plate and in your utensils, allowing you to partake in the pleasures of dining.

Step outside, and you'll witness gravity's handiwork in the swaying of trees and the rustling of leaves. The very growth of plants and the flow of water are shaped by the constant force of gravity, as they seek balance and stability in their environment.

Gravity also plays a crucial role in your daily commute. Whether you're driving a car, riding a bicycle, or taking public transportation, the forces of gravity dictate how vehicles interact with the road or tracks, enabling safe and efficient travel.

In your moments of leisure, whether you're playing sports, enjoying a swim, or practicing yoga, gravity is an ever-present partner, influencing your movements and interactions with the physical world.

This force even extends its influence beyond the Earth's surface, affecting the tides of our oceans, the phases of the Moon, and the orbits of planets in our solar system.

It's a journey to explore the everyday manifestations of this cosmic force. We'll delve into the science behind gravity, demystifying its principles in a relatable way. From the simplest actions to the grandest phenomena, we'll uncover the pervasive and intriguing role that gravity plays in shaping your world. Join us as we unveil the hidden mysteries and practical implications of gravity in your daily life, offering a new perspective on a force that is as familiar as it is extraordinary.

## 3.1 Immediate impact of gravity on Earth

Gravity's immediate impact on Earth is evident in the way it keeps us firmly grounded, ensuring that we don't float into space during our daily activities. This force influences everything from our steps on solid ground to the stability of our structures and vehicles. Additionally, gravity orchestrates the mesmerizing rise and fall of ocean tides, a phenomenon driven primarily by the gravitational pull of the Moon and, to a lesser extent, the Sun. The rhythmic dance of tides affects coastal ecosystems, maritime activities, and coastal communities. These real-world manifestations of gravity underscore its integral role in shaping our world and daily experiences, from the most ordinary actions to the captivating rhythms of nature.

**Keeping Us Grounded:**

Delve into the essence of gravity, an invisible but omnipresent force that orchestrates our everyday existence with unparalleled precision. It's a phenomenon so fundamental that it's often overlooked, yet it's the reason why we remain tethered to the Earth's surface. Imagine the simple act of standing, a seemingly mundane action. As you rise from a seated position, gravity becomes your steadfast companion, gently pulling you back to the ground. This constant interplay between you and the Earth, mediated by gravity, is what ensures you don't drift aimlessly into the cosmos.

Now, consider the act of walking—a symphony of coordinated movements involving every muscle and joint in your body. With each step, you trust that gravity will keep you grounded, providing the stability and assurance you need to move confidently through the world. Gravity is the silent partner in this dance, allowing you to maintain balance and navigate your environment effortlessly.

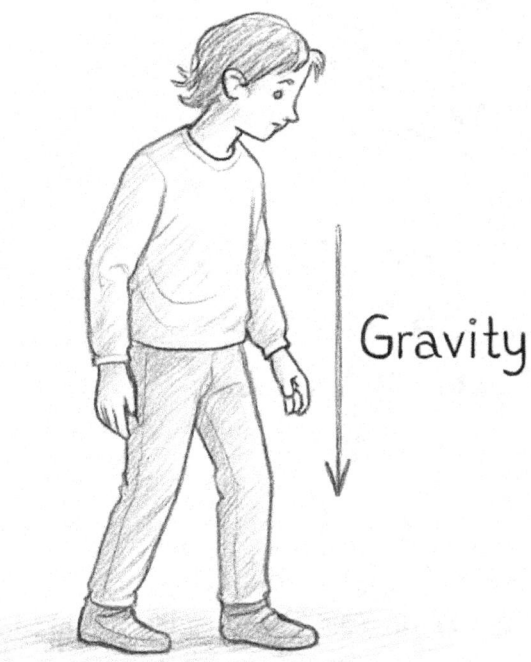

Gravity is the force that pulls us
down to the Earth's surface.

Yet, gravity's influence extends far beyond the realm of individual movements. It permeates every aspect of our daily lives, from the most basic activities to the most complex endeavors. Whether you're brewing a morning cup of coffee, driving a car, or marveling at a soaring bird in the sky, gravity is the unyielding force that governs the behavior of objects in our environment. It molds the contours of our world, ensuring that everything follows predictable trajectories under its watchful gaze. Without gravity, the very fabric of our reality would be fundamentally altered, and the world we know today would be a vastly different and more chaotic place.

## An invisible thread connects us to Earth — gravity, the quiet force that keeps every step, sip, and thought grounded in place.

In essence, gravity is the quiet architect of our existence, a force so ingrained in our daily lives that we often fail to recognize its profound significance. It is the invisible thread that weaves through the mosaic of our world, providing stability, order, and a profound sense of connection to the Earth and the cosmos beyond.

**The Tides:**

Gravity also has a dramatic effect on the Earth's oceans, creating the fascinating phenomenon of tides. The primary agents behind tides are the Moon and the Sun. The Moon's gravitational pull causes the ocean water to bulge towards it, creating high tides on the side of the Earth facing the Moon. Simultaneously, there are high tides on the opposite side, caused by the centrifugal force resulting from the Earth-Moon system's rotation.

# Gravity also has a dramatic effect on the Earth's oceans, creating the fascinating phenomenon of tides.

The Sun's gravity contributes to these tidal patterns as well, although to a lesser extent. When the gravitational forces of the Sun and the Moon align during full moons and new moons, we experience especially high tides and low tides, known as spring tides. Conversely, during the Moon's first and third quarters, when the gravitational forces oppose each other, we see lower high tides and higher low tides, known as neap tides.

Tides are not only a mesmerizing natural phenomenon but also have practical implications for coastal communities, shipping, and fishing. Understanding the gravitational dance that creates tides is essential for those who rely on the ocean for their livelihoods and for anyone who appreciates the dynamic nature of our planet.

Gravity's immediate impact on Earth is a constant and tangible presence in our lives. It's a force that keeps us grounded and shapes the behavior of the oceans, reminding us that even the most complex scientific concepts have direct and relatable effects on our daily experiences.

## 3.2 Examples like sports, weight, and travel to explain concepts

**Weight and Mass:**

Let's embark on a thought experiment that takes us to the realms of different celestial bodies, each with its unique gravitational characteristics. Picture yourself at a futuristic gym equipped not only with state-of-the-art exercise machines but also with a remarkable set of scales designed for cosmic travelers. As you step onto the scale, you're instantly transported to different worlds, each with its own gravitational pull, revealing a fascinating interplay between mass and weight.

First, imagine yourself on the Moon, Earth's closest neighbor. As you stand on the lunar surface, you notice a stark contrast in your weight compared to what you're accustomed to on Earth. The number displayed on the scale is just a fraction of your Earth weight. This dramatic reduction in weight is a direct consequence of the Moon's weaker gravitational field. You feel remarkably light, almost as if you could effortlessly leap across the lunar landscape.

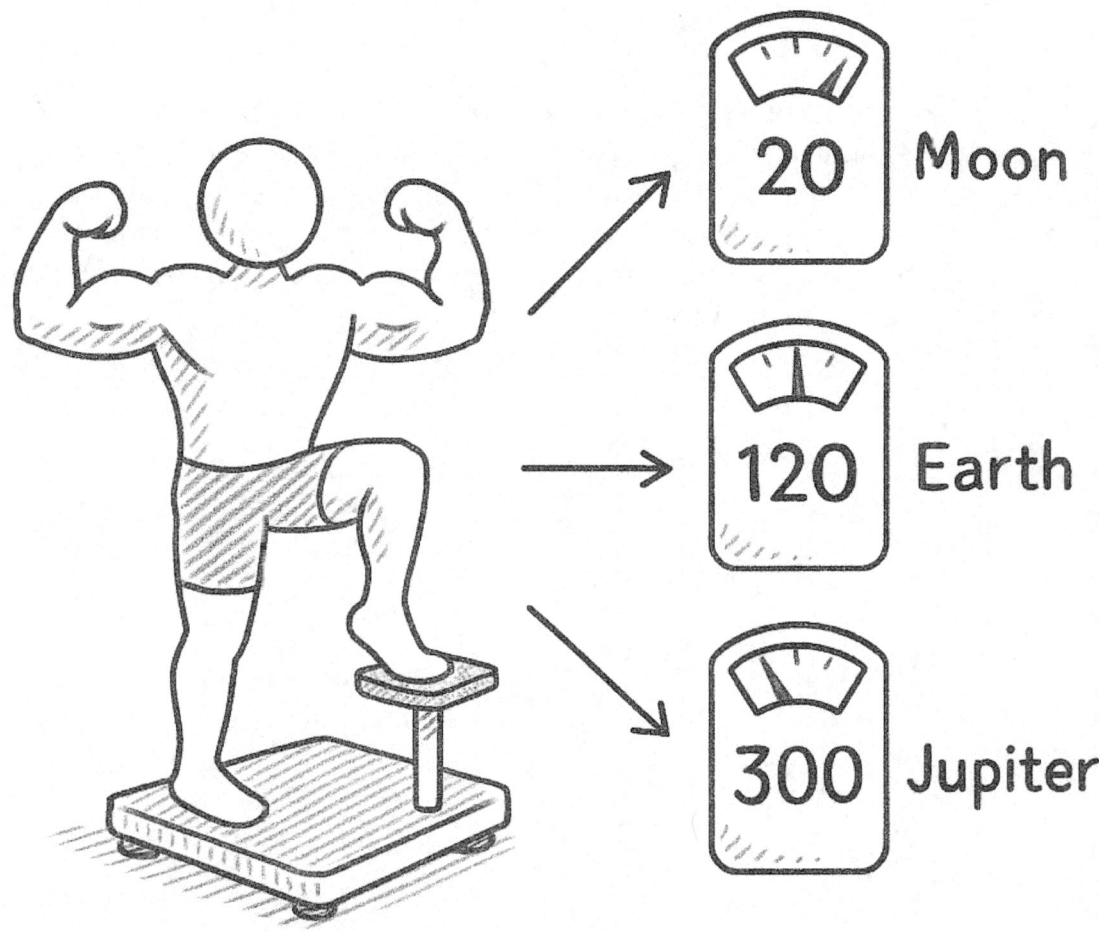

## Your weight changes across planets.

Now, let's shift our perspective to a more massive celestial body, like Jupiter, the largest planet in our solar system. As you step onto the scale on Jupiter's surface, your weight skyrockets to a magnitude far beyond your earthly experience. Jupiter's colossal size and immense gravitational force exert an overwhelming pull on your body. Every step becomes a herculean effort, and even the simplest movements require incredible strength.

Next, envision yourself on a hypothetical planet with gravity similar to Earth's but slightly stronger. On this world, you'd notice that your weight has increased compared to your familiar Earth weight. This experience underscores a crucial principle: weight depends on the local gravitational field. The stronger the gravitational pull, the greater the force pressing you against the ground, and consequently, the higher your weight registers on the scale.

In all these scenarios, one essential concept remains constant: your mass. Your mass represents the intrinsic amount of matter within your body and remains unchanged regardless of your location in the cosmos. It's a fundamental property of your being, as constant as the North Star in the night sky. This unchanging mass serves as a universal anchor that allows scientists to make comparisons and draw connections between celestial bodies, regardless of their gravitational variations.

Your weight may change across planets, but your mass — your true self — remains constant everywhere in the cosmos.

As we return from our cosmic exploration and step back onto Earth's familiar surface, we gain a deeper appreciation for the intricate interplay between mass and weight. It's a reminder that our understanding of gravity not only shapes our scientific endeavors but also enriches our imaginative journeys into the vast cosmos, where the concept of weight becomes a measure of the local gravitational embrace, and mass stands as a steadfast, unchanging companion in our cosmic adventures.

**Sports and Projectile Motion:**

Imagine yourself on a sunlit sports field, ready to engage in your favorite athletic activity, be it soccer, basketball, or any sport involving projectiles. As you prepare to make that perfect throw or kick, the invisible but ever-present force of gravity comes into play, becoming an integral part of your performance.

Take the example of a soccer match. As you line up for a powerful penalty kick, your foot connects with the ball, propelling it forward. What ensues is a ballet of physics and athleticism. The moment the ball leaves your foot, gravity asserts its influence. It doesn't pull the ball straight down but rather imparts a slight curvature to its trajectory. This behavior is due to the concept of projectile motion—a fundamental principle in physics.

As the soccer ball takes flight, it follows a graceful arc through the air. It rises, reaches its zenith, and then gracefully descends toward the goalpost. This arc is the result of gravity's constant tug, which pulls the ball downward at a uniform rate throughout its journey. Because of this, you can predict the path the ball will take with remarkable accuracy. Understanding the principles of projectile motion and gravity allows athletes to anticipate how the ball will behave in flight, enabling them to make precise calculations and decisions during fast-paced games.

In a basketball game, when a player makes a perfect jump shot, they also engage in a dance with gravity. The ball follows a similar curved path as it leaves their fingertips, arcs through the air, and either swishes through the net or bounces off the backboard or rim. Once again, gravity's influence is instrumental in determining the trajectory of the shot. This understanding of projectile motion and gravity allows basketball players to make split-second decisions about their shot angle, timing, and strength, greatly enhancing their scoring abilities.

# On the field, gravity shapes the parabolic arc of a ball in flight.

So, whether you're on a soccer pitch, a basketball court, or any sports field, the laws of physics, intertwined with gravity, come into play, creating a beautifully choreographed performance. Understanding these principles not only deepens our appreciation for the athleticism on display but also empowers athletes to hone their skills, make precise plays, and achieve those unforgettable moments of victory. It's a reminder that in the world of sports, as in the cosmos itself, gravity is an ever-present partner in the grand dance of motion and competition.

**Travel and Gravity Variations:**

Picture embarking on a breathtaking journey to a mountainous region or an elevated destination that beckons with its natural beauty. As you ascend to higher altitudes, you may begin to notice a subtle change in your physical sensations—a feeling of being slightly lighter and more buoyant than usual. This intriguing phenomenon is a direct consequence of the fundamental principles of gravity that govern our world.

When you venture to a location at a higher elevation, you are, in essence, moving away from the Earth's center, albeit by a relatively small amount. In this scenario, gravity exerts a slightly reduced gravitational force upon you compared to what you experience at lower elevations. It's a fascinating manifestation of how gravity's strength is inversely proportional to your distance from the Earth's center.

As you continue your ascent, perhaps to the summit of a majestic mountain, the sensation of being lighter becomes more pronounced. You might find that activities like climbing or hiking require less effort than they would at sea level. Even simply taking a deep breath feels refreshing, as the reduced gravitational pull makes it marginally easier for your chest muscles to expand your lungs.

Conversely, imagine a journey to a place of immense mass, such as a massive exoplanet in a distant corner of the universe. There, you would experience a strikingly different effect. The gravitational pull would be significantly stronger, and you would feel considerably heavier. Every step would be a testament to the unyielding force of gravity, demanding more effort and energy to move about.

# At higher elevations, gravity decreases slightly, making you feel lighter.

These experiences underscore the dynamic nature of gravity and how it influences our perception of weight and mass. Whether we find ourselves at higher altitudes on Earth or explore the gravitational realms of other celestial bodies, gravity serves as a constant companion, subtly shaping our interactions with the world around us. It's a reminder that the force that keeps our feet firmly planted on the ground is as dynamic and ever-changing as the landscapes we explore, continually inviting us to ponder the intricate dance between gravity and our earthly adventures.

By relating gravity to these common, relatable experiences—weight measurement, sports, and travel—we demystify this cosmic force and make it a tangible part of our lives. We come to appreciate that gravity is not an abstract concept confined to scientific textbooks but a force that shapes our physical well-being, enriches our leisure activities, and adds depth to our exploration of the world. It's a force that invites us to embrace the wonders of the universe, not as distant spectators, but as active participants in the intricate dance of gravity and existence.

# 4

# How Gravity Keeps the Universe Together

**"Gravity is the choreography of the cosmos."**
— *Brian Greene*

Gravity, the unassuming force that keeps us grounded on Earth, extends its influence far beyond our planet's boundaries, playing a central role in the vast cosmic dance of the universe. In this exploration, we uncover the captivating story of how gravity acts as the cosmic glue that binds celestial bodies, shapes galaxies, and orchestrates the cosmic symphony of the cosmos.

Imagine a galaxy, a majestic spiral of stars spinning through the cosmos. Gravity is the invisible thread that weaves these stars together, preventing them from drifting apart into the void of space. It's gravity that binds star clusters, planets, and moons to their parent stars, ensuring that they remain in their cosmic orbits, tracing elegant paths through the night sky.

Zoom out further, and we encounter galaxy clusters, vast cosmic congregations of galaxies held together by gravity's unyielding grip. These clusters are the largest structures in the universe, where the cumulative gravitational pull of countless galaxies creates a gravitational well so deep that even light itself is bent and distorted by its immense force.

But gravity's influence doesn't stop there. It's the driving force behind the formation of stars from colossal clouds of gas and dust, as well as the relentless pull that ultimately leads to the birth of black holes, enigmatic cosmic entities with gravity so intense that nothing, not even light, can escape their grasp.

In this chapter, we journey through the cosmos to explore the role of gravity in sculpting the universe's grand mosaic. We'll delve into the elegance of gravity's laws, revealing how they govern the motion of celestial bodies and shape the evolution of galaxies. We'll also ponder the cosmic mysteries that gravity unravels, from the bending of light to the existence of dark matter and dark energy.

Gravity is not just a force; it's a storyteller that reveals the hidden narratives of the cosmos. It's the cosmic glue that keeps the universe together, connecting us to the farthest reaches of space and time. As we embark on this cosmic voyage, we gain a deeper appreciation for the profound impact of gravity on the grand stage of the universe, where it continues to shape and unite all that we see and beyond.

## 4.1 The universal nature of gravity and its role in celestial bodies

Gravity is a universal force, a cosmic architect that shapes the behavior of celestial bodies throughout the vast expanse of the universe. Its influence extends far beyond Earth, touching every corner of the cosmos and playing a pivotal role in the existence and dynamics of celestial objects.

**Universal Gravitational Attraction:**

Consider the profound simplicity of gravity, a principle that extends its influence far beyond our terrestrial realm. At the heart of this cosmic force lies a fundamental truth: every object, regardless of its location in the vast expanse of the universe, exerts a gravitational pull on every other object with mass. This elegantly straightforward concept transcends the boundaries of our planet, radiating its influence throughout the cosmos.

Imagine yourself on a journey through the cosmos, traversing the moon's barren surface or gazing upon the distant flicker of a distant star. No matter where you find yourself in the celestial mosaic, you are never truly alone; gravity is your constant companion. It's the invisible thread that links you to every celestial body, binding you to the very essence of the universe itself.

# Universal Gravitational Attraction

This universal language of gravity unites all matter in the cosmos in a cosmic dance of attraction and connection. From the grandest galaxies, where stars cluster and spiral in graceful arcs, to the tiniest particles, where atoms and subatomic particles engage in intricate interactions, gravity is the unifying force that underpins the entire cosmic symphony. It reminds us that, in the grand scheme of the universe, we are all interconnected, sharing an indelible bond forged by the omnipresent embrace of gravity.

**Celestial Orbits and Kepler's Laws:**

Venture into the cosmic theater, where the celestial bodies perform an awe-inspiring ballet guided by the gentle but unwavering hand of gravity. Imagine a distant solar system, where planets glide in graceful orbits around their radiant star, tracing elliptical paths through the cosmic expanse. This cosmic choreography, described by Johannes Kepler in the 17th century through his laws of planetary motion, is one of the most captivating displays of gravity's influence on the grandest scale.

Kepler's laws reveal the intricate dance of the planets around the Sun, where gravity is the invisible conductor orchestrating each celestial movement. As you contemplate the elliptical paths traced by these planets, it becomes evident that gravity is not just a force of attraction but also the cosmic maestro that ensures celestial bodies remain tethered to their parent stars. Without this gravitational embrace, planets and other celestial objects would follow straight-line trajectories, perpetually adrift in the vastness of space, never knowing the captivating rhythms of orbit and revolution.

Zoom in on a planetary scale, and you'll find that moons are not exempt from this celestial waltz. They too are caught in the gravitational embrace of their parent planets, tracing intricate orbits that echo the larger dance of the solar system. Each moon becomes a testament to the power of gravity to forge cosmic partnerships, binding these celestial companions in eternal orbits.

Extend your gaze to the galaxies themselves, vast collections of stars, dust, and dark matter. Gravity is the cosmic glue that binds these galactic ensembles, regulating the motion of stars within their spiral arms or elliptical cores. It is the force that keeps galaxies from disintegrating into the void, holding together the cosmic mosaic of our universe.

# Celestial Orbits and Kepler's Laws

In this symphony of celestial motion, gravity is the unseen composer, crafting a cosmic masterpiece that unfolds across the canvas of the cosmos. It invites us to contemplate the intricate interplay of forces and the boundless beauty of the universe, reminding us that even the grandest celestial dances are orchestrated by the ever-present force of gravity.

**Star Formation and Stellar Evolution:**

Venture into the cosmic cradles where stars are born, and you'll find gravity at the heart of this awe-inspiring spectacle. Picture colossal clouds of gas and dust, vast and formless, drifting through the interstellar void. These cosmic reservoirs hold the potential for celestial marvels, awaiting the transformative touch of gravity. When a region within one of these clouds becomes sufficiently dense, gravity seizes control, compelling the particles to draw nearer and form a dense core.

# Gravity plays a crucial role in formation of stars from massive clouds of gas and dust.

As this gravitational embrace intensifies, something extraordinary begins to unfold—the birth of a new star. The core continues to collapse under the relentless force of gravity, and as it does so, pressure and temperature surge to unfathomable levels at its core. It is here, within this crucible of cosmic energy, that nuclear fusion ignites. This brilliant fusion reaction marks the star's official entry into the cosmic stage, illuminating the heavens with its radiant energy. Gravity has not only initiated this celestial birth but also sustains the star's existence for billions of years.

Yet, gravity's role in the stellar saga doesn't end with the birth of a star. Throughout its luminous life, gravity is the vigilant sentinel that maintains the delicate balance of cosmic forces. As nuclear fusion rages within the star's core, generating a prodigious outward pressure, gravity pushes back with equal vigor. This gravitational force acts as a cosmic referee, preventing the star from careening into catastrophic collapse under the crushing weight of its own mass or spiraling outwards in unchecked expansion.

Gravity becomes the conductor of a stellar symphony, orchestrating the harmonious interplay between the fiery core and the relentless pull of the star's mass. It's a cosmic dance that persists for eons, shaping the destiny of stars and, in turn, the destiny of entire galaxies. It is the enduring embrace of gravity that allows stars to shine brightly across the cosmic mosaic, painting the night sky with their celestial brilliance.

**Galaxies and Clusters:**

Now, let's ascend to the grand stage of the universe, where gravity plays the role of cosmic architect, sculpting the colossal structures that define the cosmos. In this vast theater of existence, galaxies emerge as the celestial protagonists, each an intricate mosaic of stars, gas, dust, and dark matter. These galaxies, with their diverse shapes and luminous inhabitants, owe their very existence to the gravitational forces that shape and mold them over the aeons.

Imagine galaxies as cosmic cities, with stars as their inhabitants, all orchestrated by gravity's invisible hand. Gravity gathers the primordial ingredients—vast clouds of gas and dust—into dense pockets within the interstellar void. These pockets, under the inexorable grip of gravity, become stellar nurseries, birthing stars by the thousands or millions. These stars, bound by gravity to their galactic home, give rise to the brilliant, intricate patterns of spiral arms and unique shapes that distinguish each galaxy.

# Gravity binds individual galaxies together into vast cosmic clusters.

But the cosmic influence of gravity doesn't stop there. Zoom out further, and we encounter galaxy clusters, the metropolises of the cosmos. These colossal gatherings of galaxies are bound together by gravity's relentless pull, creating immense gravitational fields that influence the very fabric of the universe itself. Gravity's reach extends far beyond the visible galaxies, as it exerts its influence over the dark and enigmatic matter that resides in these cosmic metropolises, shaping their dynamics and holding them together in an intricate cosmic dance.

As we gaze upon the night sky and marvel at the distant galaxies, it's gravity that bends the path of light from these celestial beacons, creating the mesmerizing cosmic lenses and distortions that astrophysicists study with keen interest. Gravity's embrace extends to the furthest reaches of the universe, influencing the distribution of matter on the grandest of scales, ultimately weaving the intricate web of cosmic structure. It's a reminder that even in the vastness of space, the force of gravity reigns supreme, etching its signature on the very fabric of the cosmos.

**The Role of Dark Matter:**

Enter the enigmatic realm of dark matter, a cosmic enigma that remains unseen but profoundly felt through its gravitational caress. Unlike ordinary matter that we encounter every day, dark matter does not engage in the dance of light; it shrouds itself in invisibility. However, its presence is undeniable, for it holds the universe's best-kept secret within its grasp.

Dark matter, though elusive in nature, is an integral cosmic player, comprising a substantial share of the universe's mass. Its existence is revealed through the unmistakable pull of gravity that it exerts on the cosmos. Imagine the universe as a vast cosmic mosaic, with galaxies and galaxy clusters woven into its intricate fabric. Dark matter is the unseen weaver, guiding the threads of matter with its gravitational threads, giving rise to the grand cosmic structure we behold.

Gravity, in its universal embrace, becomes the conduit through which dark matter exerts its influence. On the cosmic scale, dark matter forms vast halos around galaxies, acting as cosmic scaffolding that holds galaxies together and influences their rotations. It is the gravitational hand that orchestrates the intricate ballet of galaxies within galaxy clusters, sculpting the cosmic metropolises we observe through our telescopes.

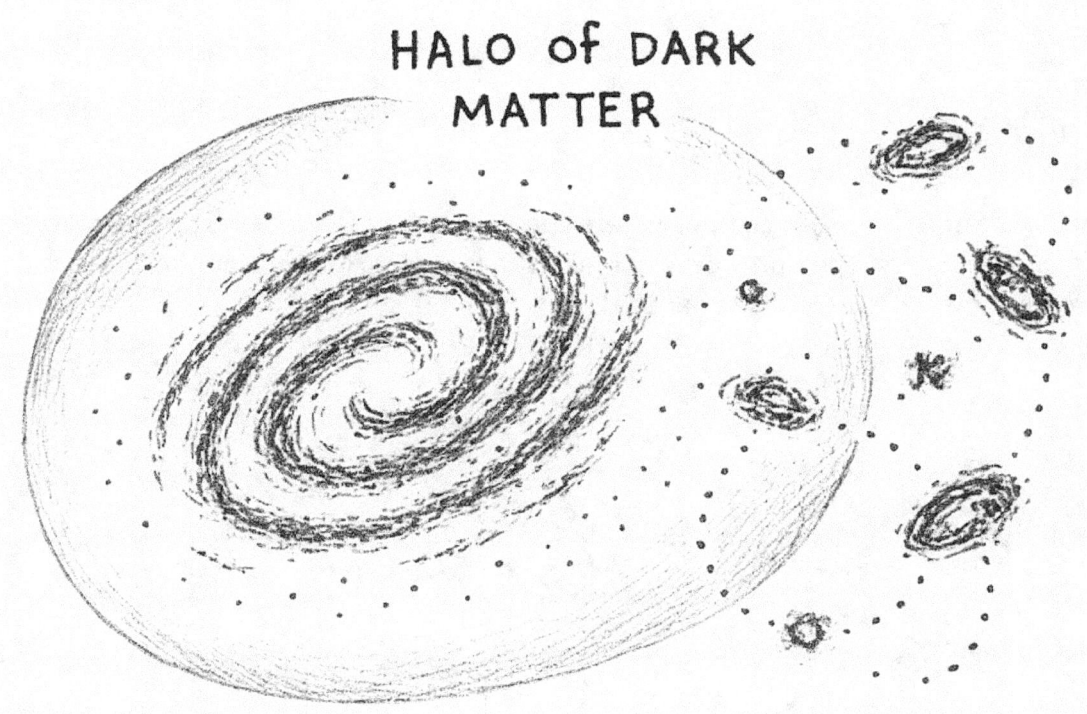

HALO of DARK
MATTER

# The gravitational effects of unseen dark matter

As we delve into the depths of the cosmos, it becomes increasingly apparent that gravity and dark matter are intertwined in a cosmic partnership, shaping the very framework of our universe. While we may not yet fully understand the nature of dark matter, its gravitational legacy serves as a profound reminder of the mysteries that continue to captivate our quest for cosmic comprehension.

In essence then, gravity is the cosmic glue that binds celestial bodies and structures together. It's the force that ensures the harmonious dance of planets, stars, and galaxies, while also revealing the hidden mysteries of the universe through its gravitational effects. Gravity's universal nature unites us with the farthest reaches of the cosmos, reminding us that we are all interconnected by this fundamental and unrelenting force that shapes the very fabric of the universe.

## 4.2 Simple analogies and visuals to explain complex concepts

Imagine a grand cosmic dance floor where the Sun is the central performer, and planets are the graceful dancers. The Sun's gravity is like the mesmerizing music that draws the planets towards it. Now, let's look at the dance steps:

## Simple analogies and visuals to explain complex concepts

## Newton's Law of Universal Gravitation

**Newton's Law of Universal Gravitation:** Visualize the elegant cosmic dance that unfolds in our solar system, where gravity orchestrates the movements of planets with the finesse of a master conductor. This celestial ballet can be likened to the simple act of twirling a ball on a string, a captivating interplay between gravity's relentless pull and the planets' stubborn inertia.

In this cosmic performance, our radiant Sun assumes the role of the hand that holds the string, while the planets take on the character of spinning balls. The Sun's gravitational force, akin to your hand's firm grip on the string, serves as the central force that beckons the planets to revolve around it. Yet, it's the planets' intrinsic inertia, their inherent determination to maintain their motion, that introduces a captivating twist to the choreography.

As the planets orbit the Sun, their inertia propels them forward, akin to the balls' desire to escape your hand's grasp and fly away. However, the gravitational pull of the Sun continually tugs at them, preventing them from wandering off into the cosmic void. This delicate balance between inertia's yearning for freedom and gravity's unyielding embrace is what gives rise to the graceful, elliptical orbits of the planets.

In this way, gravity not only keeps the planets tethered to the Sun but also ensures that they trace precise paths through the celestial expanse. It's a celestial spectacle where the fundamental laws of physics harmonize, allowing us to witness the mesmerizing celestial dances of our planetary companions as they twirl and pirouette through the cosmic stage.

**Orbits - The Cosmic Spin:** Picture our solar system as a cosmic ice rink, with the Sun as the radiant spotlight at the center, casting its gravitational glow in every direction. The planets, like skilled ice skaters, glide along the smooth surface of spacetime, following precise elliptical paths known as orbits. This celestial ice dance is a testament to the unceasing interplay between gravity's gentle tug and the planets' elegant movements.

Each planet, in its unique orbit, swirls gracefully around the Sun, mirroring the fluid motions of a seasoned skater. The Sun's gravitational influence, like the spotlight's magnetic allure, keeps the planets engaged in their perpetual performance. Just as a skater is drawn towards the captivating light of the spotlight, the planets are drawn towards the Sun by the force of gravity.

**Like skaters on an ice rink, planets glide around the Sun, held in orbit by gravity's steady pull.**

But this celestial ballet is not without its nuances. If a planet were to momentarily cease its forward motion, it would succumb to the Sun's gravitational pull and begin its descent towards the solar center. Conversely, if a planet were to accelerate to excessive speeds, it would defy gravity's grasp and embark on a journey into the cosmic void. It's the delicate balance between the planets' inertia and the Sun's gravitational hold that maintains the harmony of this cosmic ice dance, ensuring that each planet remains in its designated orbit, tracing its elegant path around our celestial centerpiece.

**Balancing Act:** Imagine a cosmic tightrope act, where planets play the role of the adept tightrope walkers, skillfully navigating the celestial equilibrium between gravity's relentless pull and inertia's unyielding resistance. This delicate cosmic tightrope dance closely mirrors the artistry of a human tightrope walker with a balancing pole.

In this cosmic spectacle, the planets' orbits become the metaphorical tightrope, stretched across the vast expanse of space. Gravity, akin to the relentless downward force that challenges the tightrope walker's poise, continuously draws the planets toward the Sun's radiant center. However, like the tightrope walker's balancing pole, inertia intervenes, propelling the planets forward with an inherent desire to maintain their cosmic stride.

# Like a tightrope walker, planets are kept in balance between the pull of gravity and the push of inertia.

Much like the tightrope walker's ever-watchful eye on their balance, the planets find their equilibrium by carefully managing the interplay between gravity's downward pull and their outward inertia. Lean too far towards the Sun, and gravity asserts its dominance, urging the planet to approach ever closer. Lean too far away, and inertia takes control, pushing the planet farther from the cosmic tightrope.

This perpetual balancing act ensures that the planets maintain their graceful orbits, never straying too close to the Sun's scorching embrace nor venturing too far into the frigid depths of space. It is this exquisite equilibrium, this cosmic tightrope walk, that paints the canvas of our solar system with the mesmerizing orbits of each planet, a testament to the enduring interplay of gravity and inertia in the grand cosmic performance.

**Cosmic Ballet:** Visualize the planets as dancers in colorful costumes, gracefully twirling around the radiant Sun. Each planet has its own unique orbit, some close to the Sun (like Mercury) and others farther away (like Jupiter). The Sun's powerful gravity serves as an invisible stage manager, ensuring that the dancers follow their choreographed paths.

# Cosmic Ballet

Imagine the Sun at the center, radiating warm light and inviting the planets to join its celestial dance. As they move, their orbits trace elegant ellipses around the Sun, akin to dancers twirling in intricate patterns. It's a cosmic ballet where gravity is the maestro, guiding the planets in their celestial performance.

This simple analogy and visual imagery help demystify the complex concept of planetary orbits. It illustrates how gravity keeps planets in check, ensuring they gracefully follow their elliptical paths around the Sun, just like dancers following the music's rhythm on a cosmic stage.

# 5

# Einstein's Gravity: The Mind-Bending Twist

*"Time and space and gravitation have no separate existence from matter."*
— *Albert Einstein*

In the realm of physics, there's perhaps no greater iconoclast than Albert Einstein, and his theory of general relativity revolutionized our understanding of gravity. In this chapter, we embark on a journey through the mind-bending twists and turns of Einstein's gravitational masterpiece.

Imagine gravity not as a force pulling objects toward one another but as a cosmic dance, where matter and energy warp the very fabric of spacetime. This is the essence of Einstein's theory. Picture a rubber sheet stretched taut, representing the fabric of spacetime. Now, place a heavy ball, representing a massive object like a planet, in the center. The ball creates a dip or a depression in the sheet, causing smaller objects, like marbles, to roll toward it. This analogy illustrates how massive objects warp spacetime, and other objects move along these curved paths. It's a visual metaphor that captures the essence of general relativity's elegant concept.

Now, consider a scene from space where light from a distant star passes near a massive object, such as a black hole. In Einstein's gravity, this light doesn't travel in a straight line but instead follows a curved path around the massive object. This phenomenon is called gravitational lensing, and it's one of the most mind-bending predictions of general relativity. Imagine looking at the night sky and seeing not only stars but also distorted images of them, like cosmic mirages. It's as if the universe itself is playing tricks on our perception, revealing the hidden influence of gravity.

Einstein's theory also predicts the existence of gravitational waves—ripples in spacetime created by the acceleration of massive objects. Think of them as cosmic vibrations, echoing through the fabric of the universe. These waves were detected for the first time in 2015, confirming a key aspect of Einstein's theory and opening a new era of gravitational wave astronomy.

**1. Warping of Spacetime:**

# Warping of Spacetime

Consider Einstein's theory of general relativity as a cosmic revelation that unveils the nature of spacetime, transforming our perception of the universe. In this groundbreaking theory, the concept of spacetime emerges—a four-dimensional mosaic that seamlessly interweaves the dimensions of space and time into a unified framework.

To grasp this profound idea, picture spacetime as a vast, flexible fabric, much like a rubber sheet stretched taut. Now, place a heavy ball on this sheet, and watch as it creates a noticeable depression, curving the fabric around it. This analogy beautifully captures the essence of general relativity: massive objects, such as planets and stars, act as celestial sculptors, molding the very fabric of spacetime.

Imagine a planet like Earth as it journeys through the cosmos. This celestial traveler follows the contours of spacetime, just as a marble would roll along the curves and slopes of the rubber sheet around the heavy ball. The spacetime dimples and warps caused by Earth's mass dictate the planet's path, keeping it in a stable orbit around the Sun. In essence, it's as if the planet is gracefully gliding along the gravitational landscape, one that's intricately shaped by the presence of massive celestial objects.

The beauty of this cosmic revelation lies in its ability to paint a vivid picture of the interconnectedness of the universe. It's a testament to the elegance of nature's design, where gravity not only binds celestial bodies in harmonious orbits but also warps the very fabric of reality itself, giving rise to the wondrous dance of the cosmos.

## 2. Bending of Light:

Gravitational lensing, a marvel of Einstein's general relativity, is nothing short of a cosmic spectacle that transforms our understanding of the universe. This phenomenon unfolds when the journey of light, emanating from a distant celestial entity like a star, takes it on a captivating detour near a massive cosmic heavyweight, be it a black hole or a galaxy cluster.

Picture yourself stargazing on a clear night, focusing your gaze on a remote star. Now, introduce a cosmic magnifying glass into the scene, an optical wonder crafted not by glass but by the immense gravitational field of a colossal galaxy. As the star's light embarks on its cosmic journey toward your eyes, it encounters this gravitational behemoth. Instead of proceeding in its customary straight-line path, the starlight gracefully curves and bends, yielding a mesmerizing transformation akin to artistry.

# The path of light from a distant star bending around a massive galaxy, depicted as a magnifying glass

In the night sky, what once was a solitary point of starlight blossoms into intricate patterns—a celestial kaleidoscope of rings, arcs, or even multiple mirrored images of the same distant star. It's as if the universe itself is an artist, painting with the hues of gravity. The observations of these phenomena, precisely as predicted by Einstein's theory, have left astronomers and stargazers alike in awe. Gravitational lensing has evolved into a powerful celestial tool, offering a unique vantage point for peering into the cosmic depths. It not only enables the study of remote and elusive objects but also unveils the enigmatic presence of dark matter, and in some instances, reveals the existence of distant exoplanets—a testament to the profound influence of gravity on the cosmos and the boundless wonders it unveils.

### 3. Detection of Gravitational Waves:

Gravitational waves, the celestial symphony of spacetime, are mesmerizing cosmic ripples forged by the intricate choreography of massive celestial bodies, such as the harmonious duet of two black holes or the graceful waltz of neutron stars locked in an orbital embrace. These waves are an extraordinary consequence of Einstein's groundbreaking theory of general relativity, which had predicted their existence, though it would take more than a century to grasp them directly.

Merging black holes emit gravitational waves

Imagine these celestial dancers, spiraling ever closer in their cosmic ballet. As they gracefully draw nearer, their immense gravitational pull whips spacetime into motion, generating waves that ripple outward at the cosmic speed limit: the speed of light. These waves possess a unique ability to stretch and squeeze spacetime itself, leaving behind a cosmic wake that echoes through the universe.

The year 2015 marked a historic milestone in our exploration of the cosmos. It was the year when scientists, armed with cutting-edge technology and the Laser Interferometer Gravitational-Wave Observatory (LIGO), achieved the seemingly impossible: they detected gravitational waves directly for the first time. Since then, these waves have graced our cosmic senses on multiple occasions, opening a new frontier in our quest to unravel the universe's secrets. They offer us a symphonic lens through which we can observe cataclysmic events such as the awe-inspiring merging of black holes and the mesmerizing collisions of neutron stars, painting a vivid picture of the cosmos in motion. Gravitational waves, much like a cosmic heartbeat, continue to reverberate across the cosmic stage, revealing previously hidden facets of the universe and bringing us closer to the melodies of the cosmos.

These mind-bending concepts represent the pinnacle of our understanding of gravity and the cosmos. They challenge our everyday intuition, revealing a universe where spacetime is flexible, light can be bent, and gravitational waves are cosmic messengers. Einstein's theory of general relativity has not only expanded our understanding of the universe but also opened up new avenues of exploration, pushing the boundaries of our knowledge beyond what was previously imaginable.

**Recapture: Einstein's theory of general relativity dramatically reshaped our understanding of the** cosmos, challenging our intuition and offering a profound new perspective on the force that shapes the universe. Here's how it accomplished this transformative feat:

Einstein's theory of general relativity ushered in a seismic shift in our understanding of the cosmos, fundamentally altering the very fabric of space and time. In this groundbreaking framework, the familiar and comforting notions of Euclidean geometry gave way to a brave new world of non-Euclidean geometry, where space and time interwove seamlessly into a single entity known as spacetime. This was no mere intellectual rearrangement; it was a revolution in the way we perceived the universe itself.

In the grand mosaic of general relativity, gravity was not a distant and mysterious force but a consequence of the cosmic stage itself. Massive objects, like stars and planets, were celestial architects, shaping the very geometry of spacetime around them. Imagine a colossal celestial body placed on a giant rubber sheet. Instead of being a passive object subject to an invisible force, the massive body became a sculptor, bending the rubber sheet to create a profound dimple in spacetime. This curvature dictated the paths of other objects nearby, causing them to follow the contours of this newly sculpted landscape. This elegant reinterpretation of gravity replaced the age-old notion of objects mysteriously attracting one another with a visual and geometric explanation that blended space and time inextricably.

This seismic shift challenged our intuition about the cosmos. Gone were the days of a simple force pulling objects together across the void of space. In its place emerged a profound understanding: objects moved along paths imprinted into the very fabric of spacetime itself. It was a revelation that demanded a leap of imagination, and it set the stage for a new era of exploration—one where the universe's secrets were unveiled not by invisible hands but by the cosmic geometry that underpinned all of reality.

Einstein's theory of general relativity revolutionized our understanding of gravity by reimagining it as the very geometry of the universe itself. In this new cosmic framework, gravity ceased to be a mysterious force acting at a distance and instead became a consequence of massive objects shaping the very fabric of spacetime.

Picture the universe as a vast, elastic canvas, where every celestial body, from the smallest pebble to the mightiest star, creates a profound impression on this canvas, like a massive weight placed upon it. This impression causes the canvas to warp and bend, forming a sort of gravitational landscape. Now, imagine objects within this landscape as marbles rolling along its contours. Their paths are not dictated by invisible forces pulling them but by the natural curves and warps in the canvas itself. This metaphor captures the essence of Einstein's profound insight: massive objects, like stars and planets, are not passive victims of gravitational forces but active participants in the sculpting of spacetime.

This reinterpretation challenged the conventional wisdom of gravity as a force of attraction and replaced it with a new geometric perspective. Objects in the universe moved not because they were pulled upon, but because they followed the curves and warps in the spacetime fabric—a realization that forever altered our perception of how the cosmos operates.

Einstein's theory of general relativity brought with it a mind-bending revelation – gravity wasn't just about the motion of objects; it could also warp the very fabric of the universe and manipulate the path of light itself. This was dramatically confirmed during a solar eclipse in 1919, which became a pivotal moment in the history of science.

Imagine standing beneath a darkened sky during a solar eclipse, peering through a telescope at a distant star. In the vicinity of the eclipsed Sun, you notice something extraordinary. The starlight doesn't travel in a straight line; instead, it appears to bend as it passes close to the massive gravitational field of the Sun. The stars you're observing aren't where they should be; they're slightly out of place, their positions subtly shifted.

This phenomenon, known as gravitational lensing, was the groundbreaking prediction of Einstein's theory. It illustrated that gravity could act like a cosmic magnifying glass, distorting the images of distant celestial objects. The 1919 solar eclipse provided empirical evidence that confirmed this prediction. It was as if gravity had unveiled a hidden power – the ability to shape not only the motion of celestial bodies but also our very perception of the cosmos. This discovery shook the foundations of physics, challenging our intuitive notions of space, time, and light. It demonstrated that the universe was far stranger and more complex than we had ever imagined.

Gravitational waves, as predicted by Albert Einstein's general theory of relativity, are a testament to the profound and unexpected ways gravity shapes the cosmos. Imagine the universe as a vast, interconnected web of spacetime, with massive celestial objects like black holes and neutron stars dancing their intricate ballet. As these massive objects revolve around each other, they generate ripples in the fabric of spacetime itself, akin to tossing a stone into a still pond.

Now, picture scientists on Earth attempting to eavesdrop on these cosmic conversations. In 2015, after decades of painstaking effort, they achieved this seemingly impossible feat. Using incredibly sensitive instruments like the Laser Interferometer Gravitational-Wave Observatory (LIGO), they detected gravitational waves for the first time, unveiling an entirely new way of observing the universe.

Gravitational waves are not like the light or electromagnetic waves we typically associate with astronomical observations. Instead, they are ripples that travel through spacetime, stretching and squeezing it as they go. The detection of these waves was an astonishing confirmation of

Einstein's theory and a powerful testament to the force of gravity. It showcased that the universe communicates its most cataclysmic events not only through the gentle whispers of light but also through the resounding echoes of gravitational vibrations. This discovery revolutionized our understanding of the cosmos, ushering in a new era of astronomy that allows us to "listen" to the universe's deepest secrets, like the collision of massive black holes or neutron stars, and explore the most extreme corners of our universe in ways previously unimaginable.

In summary, Einstein's theory of general relativity was a seismic shift in our understanding of gravity and the cosmos. It transformed gravity from a mysterious force to a fundamental aspect of spacetime geometry. It challenged our intuition by redefining how we perceive the behavior of objects under the influence of gravity, and it paved the way for groundbreaking discoveries like gravitational lensing and the detection of gravitational waves. Einstein's gravity offered us a profound new perspective on the force that shapes the universe, forever altering our comprehension of the cosmos.

## 5.2 GPS (Global Positioning System) and Time dilation

Einstein's theory of general relativity has fascinating implications for phenomena like GPS and time dilation, and we can explain these concepts in a relatable way:

**GPS and General Relativity:**

GPS (Global Positioning System) relies on a network of satellites orbiting the Earth to provide precise location and timing information. But there's a catch: these satellites are traveling at high speeds and are farther from Earth's gravitational pull, both of which affect time in ways predicted by Einstein's theory.

Imagine you have two identical clocks—one on the Earth's surface and one on a GPS satellite. According to general relativity, the clock on the satellite will tick slightly faster than the one on Earth because it experiences weaker gravity. This effect is known as gravitational time dilation. Additionally, because the satellites are moving at high speeds relative to the Earth's surface, they experience another time dilation effect, called relative time dilation. Special relativity predicts that clocks in motion tick more slowly.

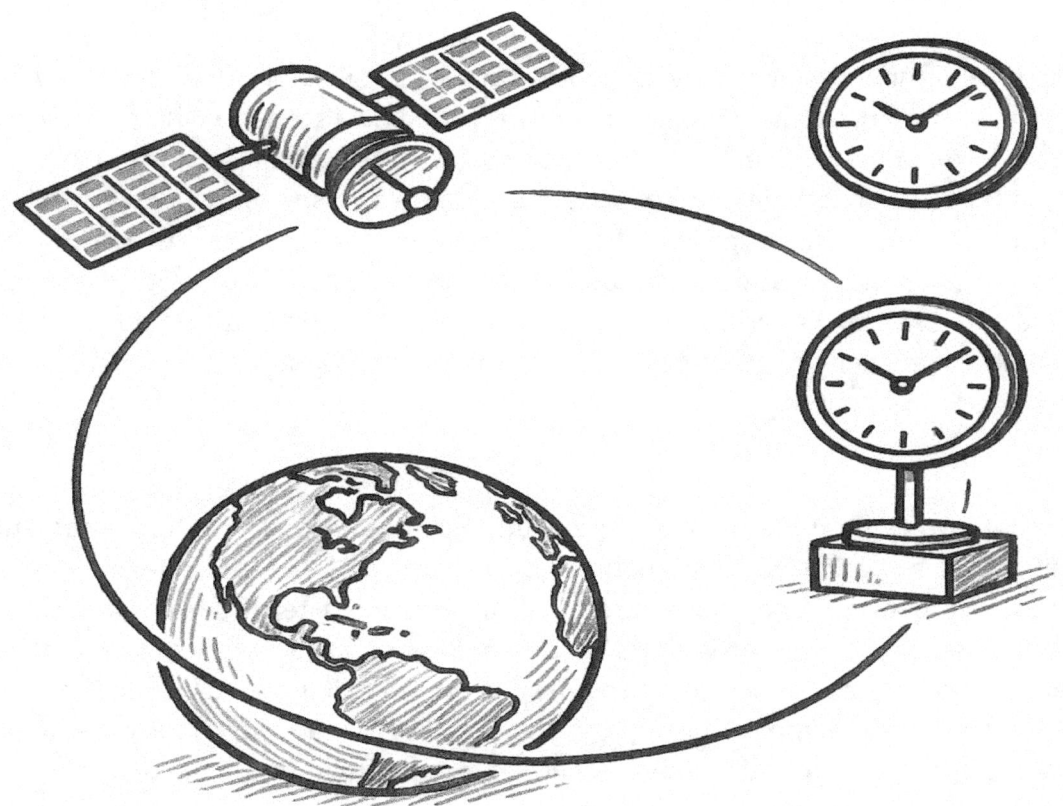

A GPS satellite carries a clock that ticks faster than an identical clock on the Earth's surface due to weaker gravity. This gravitational time dilation effect must be corrected to ensure accurate GPS measurements.

Now, let's bring this back to GPS. If we didn't correct for these time dilation effects, the satellite clocks would run faster than Earth-based clocks, leading to inaccurate GPS measurements. To account for this, engineers have to adjust the clocks on the satellites to match the time on Earth. The correction factors are incredibly tiny, but the precision of GPS requires them to be taken into account. Thus, Einstein's theory of general and special relativity plays a crucial role in the accuracy of GPS systems.

**Time Dilation and Relatable Scenarios:**

Time dilation, a mind-bending concept born from Albert Einstein's revolutionary theories, invites us to contemplate the fluidity of time itself. Imagine a scenario where you have a twin sibling, and you both decide to explore the cosmos. You board a spaceship capable of reaching speeds approaching the cosmic speed limit – the speed of light. As you journey through the vast expanse of space and return home, you are in for an astonishing surprise. While you've aged merely a few years during your interstellar adventure, your twin sibling, who remained on Earth, has aged not just years but decades. This extraordinary phenomenon isn't the plot of a science fiction novel; it's a real and profound consequence of Einstein's theory of special relativity.

The concept at play here is known as time dilation, and it arises from the fundamental idea that the speed at which you move through space affects the passage of time for you compared to someone who remains at rest. Simply put, the faster you travel, the slower time appears to pass for you. This intriguing phenomenon is the reason why astronauts aboard the International Space Station, hurtling through space at incredible speeds, experience time at a slightly slower pace than their counterparts on Earth. It's as if time itself becomes a malleable and variable entity, a realization that challenges our intuitive understanding of the steady march of time.

But time dilation isn't confined solely to the realm of high-speed space travel; it also extends its influence into the domain of gravity. Near massive celestial objects like planets or black holes, where gravitational fields are exceptionally strong, time appears to flow more slowly. Imagine embarking on a daring voyage to one of these cosmic giants, only to return and discover that less time has passed for you compared to those who remained behind on Earth. This gravitational time dilation, another intriguing facet of Einstein's theory of general relativity, demonstrates that gravity isn't just a force of attraction; it's also a cosmic timekeeper, bending and shaping the very fabric of time itself. In essence, by exploring the depths of time dilation, we peel back the layers of reality to reveal the profound interconnectedness of space, time, and the fundamental forces that govern our universe.

# Due to time dilation, the twin on Earth ages more quickly than the one traveling in space.

These examples demonstrate how Einstein's theories provide a deep understanding of how time and space behave under extreme conditions. While these effects may seem counterintuitive, they have been verified through experiments and observations and play a crucial role in our modern technological advancements like GPS and our understanding of the cosmos.

# 6

# Exploring the Cosmos: Black Holes and More

"Somewhere, something incredible is waiting to be known."
— *Carl Sagan*

In the cosmic mosaic of our universe, black holes stand as some of the most enigmatic and formidable entities. These regions of space, where gravity's grip is so intense that not even light can escape, challenge our fundamental understanding of physics and the nature of spacetime itself. This chapter takes us on a voyage into the heart of these cosmic mysteries, exploring the formation, behavior, and profound influence of black holes in the cosmos. As we peer into the abyss, we'll encounter the dazzling brilliance of supernovae, the explosive deaths of massive stars that have profound consequences for the formation of elements and the cosmic ecosystem. We'll also journey to the heart of neutron stars, where matter is crushed to densities beyond imagination, and we'll delve into the enduring mysteries of dark matter and dark energy, two enigmatic components that shape the fate of our universe. Finally, our exploration extends beyond our solar system as we seek out exoplanets, distant worlds that offer tantalizing possibilities for the existence of extraterrestrial life and expand our understanding of the cosmos. In this chapter, we navigate the cosmos, unraveling its secrets and celebrating its wonders, from the darkest depths of black holes to the farthest reaches of distant exoplanets.

## 6.1 Cosmic phenomena like black holes

In this cosmic voyage, we set our sights on an exhilarating journey through the vast expanse of the universe. As we navigate the cosmic seas, our destination is the heart of some of the most captivating and enigmatic phenomena that the cosmos has to offer. With a particular focus on the enigmatic realm of black holes, we embark on an expedition that takes us beyond the boundaries of our everyday understanding of the universe.

Our mission is to uncover the secrets hidden within the depths of space, to peer into the cosmic unknown and demystify the awe-inspiring phenomena that have both puzzled and astounded humanity for generations. We're not merely spectators; we are cosmic explorers, adventurers of the mind, and seekers of knowledge. As we venture further into this celestial odyssey, we are drawn ever closer to the cosmic wonders that await us, from the uncharted territories of black holes to the dazzling brilliance of supernovae, the staggering densities of neutron stars, and the profound mysteries of dark matter and dark energy. Each step of our journey unveils the intricate mosaic of the universe, igniting our curiosity and inspiring us to marvel at the cosmic spectacle before us.

**Black Holes: The Cosmic Mysteries**

Picture a region in the cosmic mosaic where the gravitational pull is so staggeringly intense that not even light, the fastest traveler in the universe, can escape its clutches. These enigmatic cosmic entities are known as black holes, and they stand as some of the most perplexing and awe-inspiring creations of the cosmos. In our cosmic journey, we are about to embark on an exhilarating exploration of the science, mysteries, and phenomena that surround these cosmic abysses.

Black holes, often described as the "point of no return" in space, represent a profound challenge to our understanding of physics and the very fabric of the universe. To truly grasp the enigma of black holes, we must first delve into their formation. These cosmic behemoths are born from the ashes of massive stars, the celestial giants that have exhausted their nuclear fuel and met a cataclysmic fate. When such a massive star reaches the end of its life cycle, it undergoes a dramatic implosion, collapsing under the irresistible grip of its own gravity. This gravitational collapse is akin to a stellar implosion, compressing the star's core into an infinitely dense singularity, a point where all known laws of physics seem to break down.

But black holes aren't monolithic entities; they come in various flavors and sizes. Stellar-mass black holes, for instance, can be several times more massive than our sun, while supermassive black holes, found at the centers of galaxies, can harbor millions or even billions of times the sun's mass. In our exploration, we will unravel the distinct categories of black holes, each with its unique characteristics and cosmic influence.

# A black hole is a region of space where gravity is so strong that nothing, not even light, can escape.

As we journey deeper into the mysteries of these cosmic voids, we'll also encounter the mesmerizing phenomena that dance around black holes. From the dazzling accretion disks of swirling matter to the mind-bending gravitational lensing effects that create cosmic mirages in the night sky, black holes serve as cosmic laboratories where the laws of physics are pushed to their limits. We will explore the exotic phenomena that occur in the vicinity of black holes and witness the cosmic interplay between these gravitational monsters and the surrounding universe.

So, prepare to be captivated as we embark on a voyage into the heart of darkness, where the ordinary rules of space, time, and physics cease to apply. Black holes are not merely cosmic curiosities; they represent profound puzzles that challenge our understanding of the universe and beckon us to uncover their secrets. Join us on this cosmic odyssey as we unravel the science and mysteries of these extraordinary celestial objects.

## Supernovae: Cosmic Explosions

Imagine a cosmic spectacle of unparalleled grandeur: the birth and death of stars on an unimaginable scale. At the heart of this celestial drama are supernovae, colossal explosions that stand as some of the most energetic and spectacular events in the entire cosmos. In our journey through the universe's mysteries, we will be privy to the awe-inspiring beauty and cataclysmic power of these stellar fireworks.

Supernovae are the closing act of massive stars, celestial titans that have spent their existence as cosmic beacons, illuminating the vast reaches of space. When these stellar giants reach the end of their life cycles, their destinies become inexorably linked to their own immense gravity. In a spectacular and tumultuous ballet, they undergo a dramatic implosion, followed by an explosive release of energy that outshines entire galaxies for a brief but glorious moment. These are the supernovae, the "new stars" that momentarily outshine their host galaxies and blaze across the cosmos.

But the significance of supernovae extends far beyond their breathtaking beauty. These cosmic detonations play a pivotal role in shaping the universe as we know it. As we delve into the science of supernovae, we will uncover their profound impact on the cosmos. They are the forges where elements heavier than iron are created, seeds for the formation of new stars, planets, and, ultimately, life as we know it. Supernovae scatter these elements across the cosmos, enriching the interstellar medium and providing the raw materials from which new generations of stars and planets emerge.

# A supernova is a stellar explosion that outshines an entire galaxy.

Moreover, supernovae are cosmic disruptors, triggering shockwaves that compress interstellar gas, igniting the birth of new stars. Their explosive deaths influence the dynamics of galaxies, shaping their structures and driving their evolution. These celestial explosions are not isolated events but interconnected threads in the intricate mosaic of the universe.

In our journey through the cosmos, we will stand witness to the grandeur of these stellar events, exploring their role as cosmic alchemists, stellar midwives, and architects of galaxies. Supernovae are not mere points of light in the night sky; they are the cosmic engines that drive the story of our universe, from the formation of elements to the emergence of life itself.

Join us as we unveil the beauty, science, and significance of these stellar behemoths in the grand narrative of the cosmos.

**Neutron Stars: Cosmic Densities**

Imagine a cosmic realm where the boundaries of nature are pushed to their extreme limits, where matter is compressed to densities so mind-bogglingly intense that they defy our usual understanding of the physical world. Welcome to the enigmatic domain of neutron stars, celestial marvels born from the fiery death throes of massive stars. In our cosmic odyssey, we'll embark on a journey to the heart of these stellar remnants, peeling back the layers of mystery that shroud their existence and uncovering the extraordinary physics that governs their very being.

Neutron stars are extremely dense, highly magnetized stellar remnants.

At the core of a neutron star lies a relentless battle between gravity and quantum mechanics. As the outer layers of a massive star are expelled in a dazzling supernova explosion, the core collapses under the crushing weight of gravity, collapsing inward at an astonishing rate. What emerges from this cosmic crucible is a neutron star—a sphere with a mass comparable to that of our Sun but compressed into a space barely a dozen kilometers across. This extreme density is the hallmark of neutron stars, where a teaspoon of their matter would weigh as much as a mountain on Earth.

Yet, neutron stars are not merely dense; they are cosmic dynamos. Their rapid rotations, often spinning hundreds of times per second, transform them into celestial lighthouses, emitting beams of radiation that sweep across the cosmos like the rhythmic flashes of a beacon. These pulsars, as they are known, have been instrumental in our understanding of fundamental physics and have even aided in the quest to confirm Einstein's theory of general relativity.

But the mysteries of neutron stars do not end with their incredible densities and rapid spins. These cosmic wonders also possess prodigious magnetic fields, trillions of times stronger than Earth's magnetic field. These intense magnetic forces give rise to phenomena beyond our earthly experience, from magnetar flares that outshine the entire Milky Way for a brief moment to the distortion of nearby spacetime, unveiling the mesmerizing interplay between gravity and magnetism.

Join us as we venture into the heart of these cosmic enigmas, where the laws of physics are stretched to their limits. Neutron stars are the celestial embodiment of extremes, and by delving into their secrets, we uncover not only the wonders of the cosmos but also the profound intricacies of the universe's most fundamental forces.

## Cosmic Mysteries: Dark Matter and Dark Energy

As we voyage deeper into the cosmos, we encounter two enigmatic cosmic constituents that defy our senses and stretch the boundaries of our understanding: dark matter and dark energy. These cosmic enigmas, like elusive phantoms, have captured the imaginations of astronomers and physicists alike, challenging us to peer into the abyss of the unknown and seek answers to questions that have profound implications for the very fabric of the universe.

First, let's unravel the enigma of dark matter, a substance that shrouds itself in invisibility. Although it constitutes the lion's share of the universe's mass, dark matter remains beyond the reach of our telescopes and eludes our most sensitive detectors. It neither emits, absorbs, nor reflects light, rendering it impervious to our traditional methods of observation. Yet, its presence is unmistakably felt through the gravitational dance it orchestrates on cosmic scales. Dark matter is the cosmic scaffolding upon which galaxies are built, the unseen hand that shapes the large-scale structure of the cosmos. We'll delve into the ongoing quests to detect and understand dark matter, from deep underground experiments to space-based observatories, as we strive to uncover its true nature.

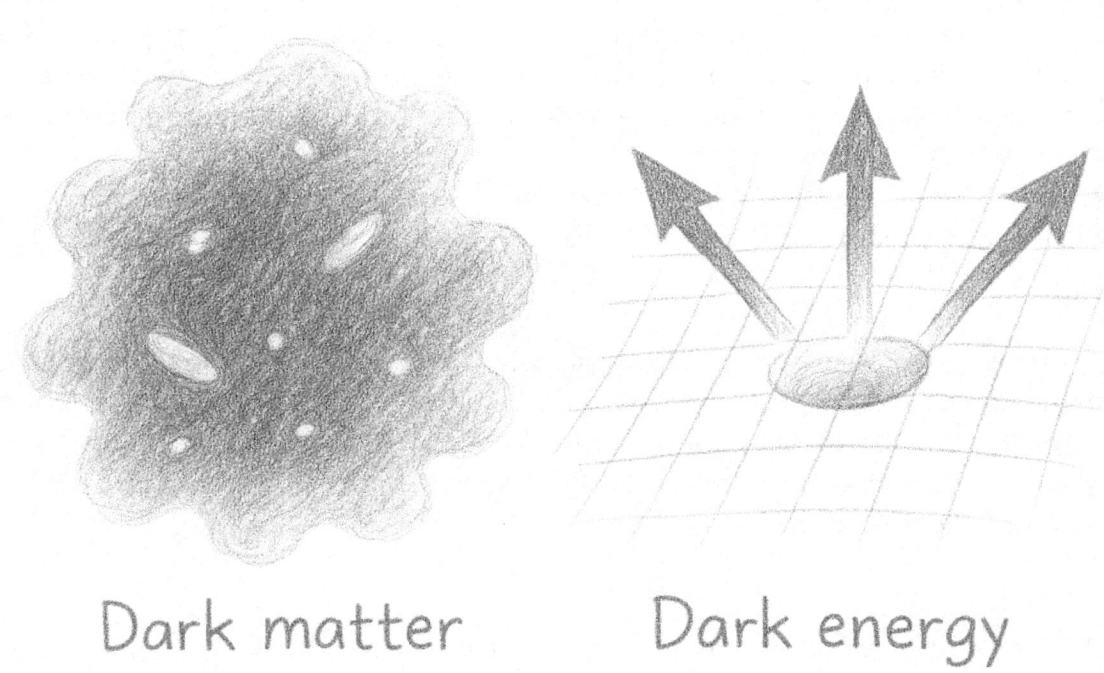

Dark matter          Dark energy

# Dark matter and dark energy are profound cosmic mysteries.

In the cosmic theater of the universe, another enigmatic actor takes center stage: dark energy. This mysterious force, unlike anything we've encountered before, is responsible for a phenomenon both astonishing and perplexing—the accelerated expansion of the universe. Dark energy, shrouded in obscurity, propels galaxies away from each other at an ever-increasing pace, defying the natural pull of gravity that seeks to draw them together. It's as if the universe carries within it a hidden engine, a cosmic repulsion that drives celestial bodies apart. We'll explore the various theories and experiments aimed at unmasking the identity of dark energy, from the subtle cosmic microwave background radiation to the powerful observations of distant supernovae.

As we journey into the depths of these cosmic mysteries, we confront questions that challenge the very essence of our understanding of the universe. What is the true nature of dark matter, and how does it interact with the known particles of the cosmos? What is the source of dark energy, and how does it shape the destiny of our universe? The pursuit of these answers not only deepens our understanding of the cosmos but also invites us to contemplate the profound interplay of forces that govern the universe's past, present, and future.

**Cosmic Exploration: Exoplanets and the Search for Life**

In our ceaseless quest to understand the cosmos, we embark on a journey that extends far beyond the boundaries of our solar system. Our destination: the enigmatic realms of exoplanets, distant worlds that dance around alien stars. These celestial wanderers, while seemingly far removed from our daily existence, hold the keys to unlocking profound insights into the nature of our universe.

Detecting exoplanets, those elusive orbs that orbit stars other than our Sun, is a scientific triumph that has unfolded in recent decades. We'll delve into the ingenious methods and instruments that have made this possible, from the precision of the transit method, where the slight dimming of a star's light reveals the presence of a planet, to the delicate wobble of a star induced by the gravitational tug of an orbiting world, which provides another crucial clue. These groundbreaking techniques have allowed us to uncover a cornucopia of exoplanets, ranging from scorching gas giants to rocky, Earth-like worlds.

But what ignites the spark of curiosity and wonder is the tantalizing possibility of finding habitable exoplanets—worlds that, like our Earth, may possess the right conditions for life to thrive. As we venture into the cosmos, we'll discuss the concept of the habitable zone, often referred to as the "Goldilocks zone," where conditions are just right for liquid water to exist, a vital ingredient for life as we know it. Exploring the potential habitability of these distant worlds, we'll consider the implications for the search for extraterrestrial life and the profound questions it raises about our place in the universe.

# Exoplanet detection and habitable worlds.

Our journey among the exoplanets is not merely a scientific endeavor but a voyage that stirs the imagination and fuels our innate curiosity. It's an exploration that invites us to contemplate the infinite diversity of the cosmos and the possibility that, somewhere among the stars, life may be flourishing on worlds yet to be discovered.

Throughout this chapter, we'll encounter the breathtaking beauty and complexity of the universe, from the extreme environments of black holes to the dazzling brilliance of supernovae. We'll also acknowledge the enduring mysteries that continue to inspire scientific exploration and fuel our curiosity about the cosmos. As we navigate the cosmos, we gain a deeper appreciation for the profound wonders and enigmas that await us in the vast expanse of the universe.

## 6.2 Important aspects of these Cosmic objects and their implications

Let's embark on an exhilarating journey to delve deeper into the mind-boggling aspects of these captivating and mysterious cosmic phenomena, along with their profound implications for our understanding of the universe. From neutron stars, where matter is compressed to unimaginable densities, to the enigmas of dark matter and dark energy that challenge our perception of the cosmos, and cosmic exploration that ignites our curiosity about the possibility of life beyond Earth, each facet of this cosmic voyage invites us to explore the universe's boundless wonders and unravel the enduring mysteries that continue to inspire awe and wonder.

### Black Holes: Cosmic Abysses

Black holes are perhaps the most mind-boggling objects in the universe. Imagine a point in space where gravity is so intense that it warps the fabric of spacetime to an extreme degree. Near a black hole, time slows down, and the laws of physics appear to break down. If you were to venture too close, you'd reach a boundary known as the event horizon—a point of no return where not even light can escape. It's as if you're peering into a cosmic abyss from which nothing emerges. The sheer gravitational force near a black hole can tear apart stars, creating brilliant flares of energy as they are devoured. Black holes challenge our understanding of the very nature of space, time, and gravity.

Black holes, those enigmatic cosmic entities, have far-reaching implications that extend beyond their mysterious, light-swallowing exteriors. These celestial objects are, in many ways, cosmic recycling centers, where the remnants of massive stars are crushed to unimaginable densities. As we journey closer to the heart of a black hole, we enter a realm where the laws of physics, as we know them, break down. The gravitational forces here are so intense that they warp the very fabric of spacetime to an extreme degree. It's a place where our understanding of the universe is put to the test.

# Black holes are perhaps the most mind-boggling objects in the universe

One of the most awe-inspiring phenomena associated with black holes is the production of gravitational waves. These are ripples in the fabric of spacetime itself, created by the acceleration of massive objects, such as two black holes or neutron stars orbiting each other. As these cosmic dancers spiral inward, they generate gravitational waves that propagate outward at the speed of light, carrying with them information about their cataclysmic interactions. Detecting these gravitational waves on Earth, as we've done with instruments like LIGO and Virgo, opens an entirely new window into the universe. It's akin to listening to the universe itself, hearing the cosmic symphony of colliding black holes and merging neutron stars. This groundbreaking field of gravitational wave astronomy allows us to explore events that were once hidden from our view, offering insights into the behavior of matter and energy under the most extreme conditions imaginable.

Black holes, in all their complexity and mystery, serve as cosmic laboratories that push the boundaries of our understanding of the universe. They're not just objects of fascination; they're gateways to unlocking the profound secrets of the cosmos, revealing the underlying laws that govern the behavior of matter and energy on the grandest scale imaginable. As we continue to probe the depths of these enigmatic cosmic phenomena, we find ourselves on a journey that reshapes our fundamental understanding of the universe itself.

## Supernovae: Explosive Stellar Deaths

Supernovae, the explosive deaths of massive stars, are cosmic phenomena that defy imagination. Consider that for a brief moment, a single star can outshine an entire galaxy. The energy released during a supernova is mind-boggling, and the explosion can scatter heavy elements like gold and uranium into space, elements that are essential for life as we know it. Supernovae are the cosmic alchemists, forging these elements and dispersing them throughout the cosmos.

The universe's grandest fireworks, supernovae, have profound implications that extend well beyond their spectacular displays. These colossal explosions are, in many ways, the cosmic engines that drive the universe's ongoing evolution. When a massive star reaches the end of its life and undergoes a supernova, it releases an astonishing amount of energy and matter into space. This expelled material is rich in heavy elements like iron, calcium, and gold—elements essential for the formation of new stars, planetary systems, and, indeed, life itself. Without supernovae, the universe would lack the raw ingredients necessary for the birth of planets and the evolution of life.

# Supernovae, the explosive deaths of massive stars, are cosmic phenomena that defy imagination.

Supernovae are also influential architects of galaxies. Their explosive deaths can influence the structure and dynamics of their host galaxies, shaping the distribution of stars and gas. In some cases, these explosions can trigger the birth of new stars as the shockwaves from the supernova compress surrounding gas clouds, prompting the formation of fresh stellar generations. Thus, supernovae are not isolated events but rather key players in the cosmic symphony, affecting the destiny of galaxies and the ongoing dance of stars within them.

Furthermore, supernovae are exquisite time capsules, allowing astronomers to study the universe's history. By observing the remnants of ancient supernovae, scientists can deduce the conditions in the early universe, providing insights into its formation and development. These celestial explosions, while brief and intense, leave behind legacies that endure for eons, leaving an indelible mark on the mosaic of the cosmos. In essence, supernovae are the cosmic storytellers, narrating tales of birth, death, and rebirth on a cosmic scale.

**Neutron Stars: Cosmic Densities**

Neutron stars are cosmic remnants of massive stars that have undergone gravitational collapse. What's truly mind-boggling is their incredible density. Imagine squeezing the mass of our Sun into a sphere the size of a city. Neutron stars are so dense that a teaspoon of their material would weigh as much as a mountain on Earth. The forces at play within a neutron star are astonishing; it's as if you've entered a realm where matter and energy exist in extreme states. Their magnetic fields are billions of times stronger than Earth's, and they can rotate incredibly fast, emitting beams of radiation like cosmic lighthouses.

Neutron stars, those incredibly dense remnants of massive stellar explosions, hold a treasure trove of scientific significance that extends far beyond their diminutive size. Their existence serves as cosmic laboratories for delving into the most extreme conditions imaginable, conditions that are simply unattainable on Earth. One of the most captivating aspects of neutron stars is their mind-boggling density. Imagine squeezing the mass of several suns into a sphere no larger than a city, and you have a neutron star. Within their core, matter is compressed to an unimaginable degree, a place where protons and electrons merge into neutrons, giving these stellar remnants their name.

These extreme conditions provide scientists with unique opportunities to probe the behavior of matter under gravitational forces that boggle the mind. The study of neutron stars has led to profound insights into fundamental physics, including our understanding of the strong nuclear force, which binds atomic nuclei together. Their intense magnetic fields, among the strongest in the universe, create environments where particles are accelerated to nearly the speed of light. This results in the generation of powerful beams of radiation, including X-rays and gamma rays, which can be detected by observatories here on Earth. By studying these emissions, scientists gain insights into the behavior of matter under the most extreme gravitational fields and magnetic pressures.

# Neutron stars are cosmic remnants of massive stars that have undergone gravitational collapse. What's truly mind-boggling is their incredible density.

Neutron stars also have a crucial role in the cosmic ecosystem. When massive stars exhaust their nuclear fuel and undergo supernova explosions, they either collapse into black holes or become neutron stars. In the process, these stellar cataclysms forge and disperse heavy elements such as gold, platinum, and uranium into the cosmos. These elements then become the building blocks for planets, asteroids, and ultimately life itself. In a sense, neutron stars are the cosmic seeders, contributing to the cosmic abundance of heavy elements that enrich the universe. Their existence challenges our understanding of matter and gravity and continues to inspire scientists to unlock the mysteries of the universe's most extreme environments.

**Dark Matter and Dark Energy: Cosmic Enigmas**

Dark matter and dark energy are cosmic enigmas that defy our senses and understanding. Dark matter makes up the majority of the matter in the universe but doesn't interact with light or any known forces, rendering it invisible. Dark energy is a mysterious force that is causing the universe's expansion to accelerate, pushing galaxies apart at an ever-increasing rate. The mind-boggling aspect is that we can't see, touch, or directly detect these entities, yet they shape the fate and structure of the cosmos.

Dark matter and dark energy are cosmic enigmas that defy our senses and understanding.

The implications of dark matter and dark energy reach deep into the fabric of the cosmos, fundamentally reshaping our understanding of the universe's structure and fate. Dark matter, although invisible and enigmatic, plays a pivotal role in the cosmic ballet of galaxies. While we cannot directly observe dark matter, its presence is inferred through its gravitational effects on visible matter, such as stars and galaxies. It forms a cosmic scaffold, providing the gravitational glue that holds galaxies together. Without dark matter, galaxies would not have enough mass to remain coherent, and the universe would lack the rich mosaic of cosmic structures we observe today.

Beyond the gravitational realm of dark matter lies the enigmatic realm of dark energy. Unlike dark matter, which pulls matter together, dark energy pushes the universe apart. It's the mysterious force responsible for the universe's accelerating expansion, a revelation that defied our expectations and transformed cosmology. Dark energy dominates the large-scale structure of the cosmos and shapes its destiny. Understanding its nature is one of the most profound challenges in modern physics, as it forces us to reconsider our understanding of fundamental forces and the very fabric of space and time.

These cosmic mysteries beckon scientists and astronomers to embark on quests to unveil their secrets. They challenge our comprehension of the universe at its grandest scales, from the behaviors of galaxies and galaxy clusters to the ultimate fate of our cosmos. The study of dark matter and dark energy has opened new frontiers in physics, promising to provide insights that will not only deepen our understanding of the universe but may also revolutionize our perception of the fundamental forces that govern it. The pursuit of these mysteries is a testament to the enduring human spirit of curiosity and exploration.

**Cosmic Exploration: Exoplanets and the Search for Life**

Cosmic exploration takes us beyond the confines of our own planet and into the vastness of the cosmos. Imagine peering through telescopes and detecting planets orbiting distant stars, known as exoplanets. Some of these worlds may reside in the "Goldilocks zone," where conditions are just right for liquid water and potentially life to exist. The sheer scale of cosmic exploration challenges our perception of distance and possibility. It's as if we're explorers setting sail on a cosmic sea, searching for new lands and, perhaps, new life forms.

The implications of cosmic exploration extend far beyond the confines of our home planet. When we venture beyond Earth's boundaries to explore the cosmos, we embark on a journey that carries profound implications for our understanding of the universe and our place within it. One of the most captivating aspects of cosmic exploration is the discovery of exoplanets, planets orbiting distant stars. This expanding roster of exoplanets unveils the incredible diversity of planetary systems throughout the universe. Each new discovery challenges our preconceptions about what a planetary system can be like and raises tantalizing questions about the potential for life beyond Earth.

# Cosmic exploration takes us beyond the confines of our own planet an into the vastness of the cosmos.

Moreover, the search for exoplanets has kindled our curiosity about the possibility of extraterrestrial life. As we find more exoplanets residing in the habitable zones of their stars, where conditions may be suitable for liquid water and life as we know it, the prospect of discovering signs of life elsewhere in the cosmos becomes increasingly tantalizing. It prompts us to contemplate the profound implications of such a discovery, from its impact on our understanding of life's place in the universe to the potential for contact with intelligent civilizations.

Cosmic exploration also serves as a crucible for technological innovation. The challenges of traveling through space, conducting scientific experiments in extreme environments, and communicating across vast cosmic distances push the boundaries of what humanity can achieve. These technological advancements, born out of our desire to explore and understand the cosmos, often find applications back on Earth, benefiting society in ways we might not have anticipated. From healthcare to communications to materials science, the innovations driven by space exploration have the potential to transform our daily lives.

In essence, cosmic exploration is a testament to our insatiable curiosity and our enduring quest for knowledge. It invites us to look beyond our terrestrial boundaries, inspiring us to dream of distant worlds and the mysteries they hold. It reminds us that the universe is a vast and wondrous place, waiting for us to unlock its secrets through the spirit of exploration and discovery.

In each of these cosmic phenomena and explorations, we encounter the extraordinary and the inexplicable, pushing the frontiers of our knowledge and inspiring awe and wonder. They beckon us to delve deeper into the mysteries of the universe, reminding us of the boundless wonders that await our exploration and discovery.

# 7

# Your Gravity Questions Answered

"I think nature's imagination is so much greater than man's,

she's never going to let us relax."

— Richard Feynman

In this section, we'll address some of the most common and intriguing questions about gravity, offering explanations and insights into this fundamental force that shapes our world. Whether you've wondered why objects fall or how gravity influences time, these answers will unravel the mysteries of gravity and its role in the universe.

## 7.1 Gravity Q&A

**Q:** What is gravity?

**A:** Gravity is a fundamental force of nature that causes objects with mass or energy to be attracted to each other. It's what keeps us anchored to the Earth and governs the motion of planets, stars, and galaxies.

**Q:** Why do objects fall to the ground?

**A:** Objects fall to the ground due to gravity's pull. When you drop something, like a ball, gravity exerts a force on it, causing it to accelerate toward the Earth's center.

**Q:** How does gravity work in space?

**A:** Gravity works the same way in space as it does on Earth. Any two objects with mass in the universe attract each other gravitationally. In space, there's just less air resistance to slow things down.

**Q:** Why do astronauts float in space?

**A:** Astronauts float in space because they're in a state of constant freefall around the Earth. The spacecraft they're in is also falling, so everything inside appears weightless.

**Q:** Does gravity affect time?

**A:** Yes, gravity affects time. According to Einstein's theory of general relativity, strong gravitational fields, like those near massive objects, cause time to pass more slowly. This phenomenon is known as time dilation.

**Q:** Can we escape Earth's gravity?

**A:** To completely escape Earth's gravity, you'd need to reach a speed of about 25,000 miles per hour (40,000 kilometers per hour). This is called escape velocity. Rockets achieve this speed to break free from Earth's gravitational pull.

**Q:** How does gravity hold the solar system together?

**A:** Gravity holds the solar system together by keeping planets in orbit around the Sun. The Sun's immense gravity creates a gravitational pull that keeps planets like Earth in their elliptical paths.

**Q:** What's the difference between weight and mass?

**A:** Mass is the amount of matter in an object and is measured in kilograms or pounds. Weight, on the other hand, is the force of gravity acting on an object's mass and is measured in newtons or pounds-force.

**Q:** Can we manipulate gravity?

**A:** Currently, we have no technology to manipulate gravity. It remains one of the least understood forces in physics. In science fiction, concepts like anti-gravity are explored, but in reality, gravity manipulation remains a challenge.

Gravity: unraveling the mysteries of the cosmos

**Q:** Why do objects of different masses fall at the same rate on Earth?

**A:** Objects of different masses fall at the same rate on Earth due to the principle of universal gravitation, discovered by Isaac Newton. This law states that all objects, regardless of their mass, experience the same acceleration due to gravity near the surface of the Earth, which is about 9.81 meters per second squared (32.2 feet per second squared).

**Q:** Can gravity be shielded or blocked?

**A:** Gravity cannot be easily shielded or blocked with conventional materials. It's a fundamental force that acts at a long range. Massive objects like planets or large spacecraft can create gravitational fields, but everyday materials do not significantly affect gravity.

**Q:** Does gravity exist everywhere in the universe?

**A:** Yes, gravity exists everywhere in the universe where there is mass or energy. However, its strength varies depending on the amount of mass and the distance between objects. In regions with very low mass density, such as deep space, gravity's effects can be extremely weak.

**Q:** How does gravity affect the shape of celestial bodies?

**A:** Gravity plays a crucial role in shaping celestial bodies. For planets, stars, and galaxies, gravity pulls matter together into spherical or nearly spherical shapes. On smaller scales, like asteroids, gravity may result in irregular shapes.

**Q:** Can you feel the effects of gravity from other celestial bodies, like the Moon or Mars, while on Earth?

**A:** While you can't feel the effects of the Moon's or Mars' gravity directly, you do experience the overall gravitational force from Earth, which includes contributions from all nearby celestial bodies. However, the effect of these distant bodies is extremely small compared to Earth's gravity.

**Q:** How does gravity influence the motion of satellites and space probes?

**A:** Gravity is the force that keeps satellites and space probes in orbit around celestial bodies like Earth or other planets. It provides the necessary centripetal force to counteract the tendency of these objects to move in a straight line and allows them to stay in stable orbits.

**Q:** Can gravity be harnessed for energy generation?

**A:** Gravity can be harnessed for energy generation through technologies like hydroelectric power, where the gravitational potential energy of water is converted into electricity. Additionally, concepts like tidal power use the gravitational interaction between celestial bodies, such as the Moon and Earth, to generate energy from the tides.

**Q:** How does gravity affect the flow of time in the universe?

**A:** According to Einstein's theory of general relativity, gravity affects the flow of time. In strong gravitational fields, time passes more slowly compared to weaker gravitational fields. This effect, known as gravitational time dilation, has practical applications, such as the adjustment of GPS satellite clocks to account for their relative motion and gravitational differences.

**Q:** Why does the Moon have lower gravity than Earth?

**A:** The Moon has lower gravity than Earth because it has less mass. Gravity is directly proportional to mass, so celestial bodies with greater mass, like Earth, exert stronger gravitational forces.

**Q:** What happens to gravity in space, far from any massive objects?

**A:** In space, far from any massive objects, gravity weakens with distance. The farther you are from a massive object like a planet or star, the weaker the gravitational force becomes. However, gravity is never entirely absent; it extends infinitely into space.

**Q:** Can gravity affect light?

**A:** Yes, gravity can affect light. According to general relativity, massive objects can bend the path of light, causing gravitational lensing. This effect has been observed and verified during astronomical observations.

# The wonders of gravity

**Q:** Why do astronauts experience weightlessness in space?

**A:** Astronauts experience weightlessness in space because they are in constant freefall around the Earth or another celestial body. The spacecraft and everything inside it are falling toward the planet due to gravity, but they are moving forward at a sufficient speed to maintain a stable orbit. This creates the sensation of weightlessness.

**Q:** Can you feel gravity from distant celestial bodies, like stars?

**A:** While you can't feel the gravitational pull from distant celestial bodies like stars, their gravitational influence contributes to the overall gravitational field of the galaxy or solar system. However, the gravitational force from individual stars is typically very weak unless you are extremely close to them.

**Q:** Does gravity affect the behavior of galaxies?

**A:** Yes, gravity plays a crucial role in the behavior of galaxies. It keeps stars and other objects within galaxies in orbit around their centers. Additionally, the gravitational interactions between galaxies can lead to events like galaxy collisions and mergers.

**Q:** How does gravity relate to the curvature of spacetime?

**A:** According to Einstein's theory of general relativity, gravity is the result of the curvature of spacetime caused by the presence of mass and energy. Massive objects, like planets and stars, warp the fabric of spacetime around them, and other objects move along curved paths in response to this curvature, creating the effect we perceive as gravity.

**Q:** Are there any hypothetical scenarios where gravity could behave differently from what we observe?

**A:** While our current understanding of gravity is supported by extensive experimental evidence, scientists continue to explore its behavior in extreme conditions, such as at the centers of black holes or during the early moments of the universe's existence. In these contexts, gravity may exhibit behaviors that challenge our current theories.

**Q:** Can gravity ever become repulsive, like a "gravitational repulsion" force?

**A:** In our current understanding of physics, gravity is always attractive, meaning it pulls objects together. However, the concept of "dark energy" is a form of energy that has a repulsive gravitational effect, leading to the accelerated expansion of the universe. It's important to note that dark energy behaves differently from the gravity we experience on smaller scales.

**Q:** What would happen if Earth's gravity suddenly disappeared?

**A:** If Earth's gravity were to suddenly disappear, everything on the planet would no longer be held to its surface. Objects, including people, buildings, and the atmosphere, would float off into space. It would be a catastrophic event with dire consequences for life on Earth.

**Q:** How does gravity influence the formation of galaxies and galaxy clusters?

**A:** Gravity is the driving force behind the formation of galaxies and galaxy clusters. It attracts matter, including gas, dust, and dark matter, causing them to clump together under its gravitational pull. Over billions of years, these clumps of matter grow, forming galaxies and clusters of galaxies.

**Q:** Can gravity affect the behavior of light in a strong gravitational field, like near a black hole?

**A:** Yes, near massive objects like black holes, gravity can significantly affect the behavior of light. The strong gravitational field bends the path of light, causing a phenomenon called gravitational lensing. It can distort or magnify the appearance of distant objects as their light travels through the gravitational field.

**Q:** Is there a connection between gravity and the expansion of the universe?

**A:** Yes, there is a connection between gravity and the expansion of the universe. While gravity acts to pull objects together, the presence of dark energy, which has a repulsive gravitational effect, is responsible for the accelerated expansion of the universe. The balance between these two forces determines the overall fate of the cosmos.

**Q:** Can objects be in a state of microgravity on Earth?

**A:** Yes, objects can experience a state of microgravity on Earth, often in the context of parabolic flight paths or aboard spacecraft in low Earth orbit. During such instances, objects and people inside the aircraft or spacecraft appear to be weightless because they are in freefall, counteracting the effects of gravity.

Gravitational waves open up new ways of learning about the universe.

**Q:** How do scientists study the effects of gravity on living organisms, like plants and animals, in space?

**A:** Scientists study the effects of gravity on living organisms in space by conducting experiments aboard the International Space Station (ISS) and other space missions. These experiments help us understand how microgravity affects various biological processes, from cell growth to bone density, and have implications for human health and space travel.

**Q:** Can gravitational waves provide new insights into the universe?

**A:** Absolutely. Gravitational waves, ripples in spacetime caused by the acceleration of massive objects, offer a new way to explore the universe. They allow us to observe cataclysmic events like the collision of black holes and neutron stars, providing valuable data that enhances our understanding of gravity and the cosmos.

**Q:** How does gravity influence the behavior of light around massive objects like black holes?

**A:** Gravity can bend the path of light around massive objects in a phenomenon known as gravitational lensing. Near a black hole, the intense gravitational field can cause light from distant objects to follow curved trajectories, creating optical distortions and multiple images of the same object.

**Q:** Can gravity vary in strength from one location in the universe to another?

**A:** Yes, gravity's strength can vary depending on the distribution of mass and energy in the universe. In regions with more mass, such as near massive galaxies or galaxy clusters, gravity is stronger. In cosmic voids, where matter is sparse, gravity is weaker.

**Q:** Are there other forces besides gravity that can influence the motion of celestial objects?

**A:** Yes, celestial objects can be influenced by forces other than gravity. For instance, electromagnetic forces, such as radiation pressure from stars or interactions with magnetic fields, can impact the motion of objects in space. Additionally, cosmic objects can experience tidal forces from nearby massive bodies.

**Q:** How do scientists detect and study gravitational waves?

**A:** Scientists detect gravitational waves using specialized instruments known as interferometers, like LIGO and Virgo. These instruments measure tiny fluctuations in spacetime caused by gravitational waves passing through the Earth. Gravitational wave detectors have provided groundbreaking insights into cosmic events like black hole mergers.

**Q:** Does gravity have an effect on the behavior of subatomic particles, like electrons or quarks?

**A:** Yes, gravity does have an effect on subatomic particles, but it is extremely weak at the subatomic scale compared to other fundamental forces, like electromagnetism and the strong and weak nuclear forces. Gravity's influence on subatomic particles becomes more significant only in extremely massive objects, like neutron stars or black holes.

**Q:** Can gravity be quantized, like other fundamental forces in quantum physics?

**A:** The quest to reconcile gravity with quantum physics, known as quantum gravity, is ongoing. While the other fundamental forces have been successfully described using quantum field theory, gravity's quantization remains a challenging problem in theoretical physics. The search for a consistent theory of quantum gravity is a major goal of modern physics.

**Q:** How does gravity influence the structure and evolution of the universe on a cosmic scale?

**A:** Gravity plays a pivotal role in the large-scale structure and evolution of the universe. It causes matter to clump together into galaxies and galaxy clusters, and it drives the expansion of the universe. The interplay between gravity, dark matter, and dark energy determines the fate and geometry of the cosmos.

**Q:** Can gravitational anomalies, where gravity behaves unexpectedly, occur in the universe?

**A:** Gravitational anomalies are a subject of scientific interest. While gravity's behavior is generally well-understood within the framework of general relativity, there may be unexplored phenomena, such as exotic matter or interactions beyond our current understanding, that could lead to gravitational anomalies under extreme conditions.

**Q:** Are there any theories or experiments that suggest gravity might not be a fundamental force but rather an emergent property of something else?

**A:** Some theories in theoretical physics, such as loop quantum gravity and emergent gravity models, propose that gravity could emerge from more fundamental quantum phenomena. These theories aim to unify gravity with other fundamental forces, like electromagnetism, within a single framework. While these ideas are still under active investigation, they offer intriguing possibilities.

Gravitational effects stretch across the cosmos, inspiring us with awe.

**Q:** Can gravity waves interact with matter, similar to how electromagnetic waves interact with charged particles?

**A:** Gravitational waves primarily interact with spacetime itself, causing ripples in the fabric of the universe. Unlike electromagnetic waves, which interact with charged particles, gravitational waves pass through matter largely unaffected. However, extremely intense gravitational waves from catastrophic events, like neutron star mergers, can potentially have subtle effects on matter.

**Q:** How does gravity influence the behavior of cosmic structures, such as galaxy clusters and superclusters?

**A:** Gravity is the dominant force responsible for the formation and evolution of cosmic structures. It attracts matter, including dark matter, gas, and galaxies, into massive structures like galaxy clusters and superclusters. Over cosmic timescales, gravity causes these structures to grow and merge.

**Q:** Can gravity be shielded or negated by advanced technologies, like science fiction force fields or anti-gravity devices?

**A:** In the realm of science fiction, various concepts explore the idea of shielding or negating gravity using advanced technologies. However, in our current understanding of physics, gravity remains a fundamental force that is extremely challenging to manipulate or negate with technology.

**Q:** How does gravity relate to the concept of a unified theory of physics, which aims to explain all fundamental forces within a single framework?

**A:** Gravity's role in a unified theory of physics is a key challenge in theoretical physics. Such a theory, often referred to as the theory of everything, seeks to unify gravity with the other fundamental forces, like electromagnetism, the strong nuclear force, and the weak nuclear force. Achieving this unity could provide a deeper understanding of the fundamental laws of the universe.

**Q:** Are there any experiments or observations that have tested the behavior of gravity in extreme environments, such as black holes or the early universe?

**A:** Scientists have conducted various experiments and observations to test gravity's behavior in extreme environments. Examples include studying the orbits of stars near supermassive black holes and observing the cosmic microwave background radiation to understand the universe's early evolution. These studies contribute to our understanding of gravity's behavior in extreme contexts.

**Q:** How does the study of gravity impact our understanding of cosmology, the study of the overall structure and evolution of the universe?

**A:** Gravity is fundamental to the field of cosmology. It shapes the large-scale structure of the universe, influences the expansion rate of the cosmos, and determines its ultimate fate. The study of gravity's role in cosmology allows us to explore questions about the universe's origin, composition, and destiny.

**Q:** Can gravity affect the flow of time on a cosmic scale, such as the age of the universe?

**A:** Yes, gravity can indeed affect the flow of time on a cosmic scale. According to general relativity, strong gravitational fields, such as those near massive celestial objects or within the universe itself, can cause time to pass more slowly. This time dilation effect has implications for the age of the universe and its expansion history.

**Q:** Are there any theoretical concepts that suggest the existence of exotic forms of matter or energy that could influence gravity differently from normal matter?

**A:** Some theoretical concepts, such as exotic matter with negative mass or energy, have been proposed in the context of wormholes or warp drives. These exotic forms of matter could potentially influence gravity in ways that differ from normal matter. However, their existence remains speculative, and they have not been observed.

**Q:** Can gravity affect the behavior of particles at the quantum level, and if so, how does this relate to quantum gravity theories?

**A:** Gravity's influence at the quantum level is extremely weak compared to the other fundamental forces, which makes it challenging to study within the framework of quantum mechanics. The quest for a theory of quantum gravity aims to reconcile gravity with quantum physics and understand how gravity behaves on the smallest scales, where classical physics breaks down.

GRAVITY BINDS THE COSMOS
AND BENDS TIME—INVITING
US TO QUESTION EVERYTHING.

**Q:** How do gravitational waves provide information about the inner workings of massive celestial objects, like black holes and neutron stars?

**A:** Gravitational waves carry information about the motion and properties of massive celestial objects. When black holes or neutron stars merge, they emit gravitational waves that can be detected by observatories like LIGO and Virgo. The waveform of these waves provides details about the objects' masses, spins, and the spacetime around them.

**Q:** Can gravity ever lead to the formation of structures or phenomena that defy our current understanding of physics, such as traversable wormholes or time loops?

**A:** Gravity's influence can indeed lead to the formation of structures like black holes, which are governed by extreme gravitational forces. Theoretical physics explores the possibilities of exotic phenomena, such as traversable wormholes or time loops, but these concepts often rely on speculative forms of matter or energy and have not been observed or confirmed.

**Q:** How does gravity interact with the fabric of spacetime, and how does this relate to Einstein's theory of general relativity?

**A:** According to Einstein's theory of general relativity, gravity is not a force acting across space but a result of the warping of spacetime by mass and energy. Massive objects, like planets and stars, create "dips" or curvatures in spacetime, and other objects move along curved paths within this curved spacetime. The presence of mass and energy tells spacetime how to curve, and the curvature tells objects how to move.

**Q:** Are there any experiments or missions planned for the future that will further test or expand our understanding of gravity?

**A:** Yes, numerous experiments and missions are planned or ongoing to advance our understanding of gravity. These include space-based missions to study gravitational waves, experiments to test gravity's behavior at small scales, and astrophysical observations to explore the effects of gravity on cosmic scales. Future discoveries in these areas promise to deepen our understanding of this fundamental force.

**Q:** How does gravity impact the stability and dynamics of planetary systems, such as our solar system?

**A:** Gravity plays a central role in the stability and dynamics of planetary systems. It determines the orbits of planets, moons, and other celestial objects. In our solar system, the gravitational pull of the Sun keeps planets in their orbits and influences the structure of the system.

**Q:** Can gravity assist space missions, such as spacecraft traveling to distant planets or exploring the outer solar system?

**A:** Yes, gravity assists, also known as gravitational slingshots or gravity boosts, are techniques used in space missions to harness the gravity of planets or other celestial bodies to alter a spacecraft's trajectory and increase its speed. These assists are crucial for efficient interplanetary travel and exploration.

**Q:** How does the presence of dark matter in the universe affect our understanding of gravity and its role in cosmic structures?

**A:** The presence of dark matter in the universe has significant implications for our understanding of gravity. Dark matter is believed to be a form of matter that does not emit or interact with light, but it exerts gravitational influence. It plays a crucial role in the formation of large-scale cosmic structures like galaxy clusters and helps explain observed gravitational effects that would otherwise be unaccounted for.

**Q:** Can gravity waves provide insights into the early moments of the universe, such as the Big Bang?

**A:** Gravitational waves have the potential to provide insights into the early moments of the universe. While they primarily result from cataclysmic events like black hole mergers, the detection of primordial gravitational waves—ripples in spacetime from the Big Bang itself—remains a goal of cosmological research. Such waves could offer direct evidence of cosmic inflation, an early rapid expansion of the universe.

**Q:** How do the orbits of natural and artificial satellites illustrate the principles of gravity and the stability of orbital mechanics?

**A:** The orbits of natural satellites (moons) and artificial satellites, including those used for communication and Earth observation, are governed by the principles of gravity and orbital mechanics. These orbits follow elliptical paths, where the gravitational pull of the central body (e.g., Earth) is balanced by the satellite's velocity. Understanding these principles is crucial for the design and operation of satellite missions.

Gravity teaches us to question, explore, and reach for the stars.

**Q:** Are there any intriguing phenomena or mysteries related to gravity that scientists are actively investigating today?

**A:** Yes, scientists are actively investigating several intriguing phenomena related to gravity. These include the nature of dark energy, the behavior of gravity in extreme environments like black holes, the search for deviations from general relativity, and the quest for a consistent theory of quantum gravity. These frontiers of research hold the promise of expanding our understanding of gravity and its role in the universe.

**Q:** Can the study of gravity contribute to advancements in space exploration, such as missions to other planets or interstellar travel?

**A:** The study of gravity is crucial for advancing space exploration. It informs mission planning, trajectory calculations, and spacecraft design. Understanding gravity's effects on celestial bodies is essential for safe and successful missions to other planets, asteroids, and even for long-term goals like interstellar travel.

**Q:** How does gravity relate to the formation and evolution of stars, including processes like nuclear fusion in their cores?

**A:** Gravity is fundamental to the formation and evolution of stars. It causes the collapse of gas and dust clouds into protostars, initiating nuclear fusion in their cores. The balance between gravity's inward pull and the pressure from nuclear reactions determines a star's stability and eventual fate, which can range from becoming a stable star like our Sun to undergoing supernova explosions or collapsing into a black hole.

**Q:** How does the concept of gravity connect with the search for habitable exoplanets and the possibility of extraterrestrial life?

**A:** The study of gravity is essential in the search for habitable exoplanets and extraterrestrial life. It helps scientists determine a planet's potential to retain an atmosphere, which is critical for sustaining life. Understanding a planet's gravitational pull also informs calculations of surface conditions, like temperature and pressure, which are vital factors for habitability.

**Q:** Can gravitational waves reveal information about the composition and structure of celestial objects, such as neutron stars or the cores of gas giants?

**A:** Gravitational waves can indeed reveal information about the composition and structure of celestial objects. By analyzing the gravitational wave signals generated by events like neutron star mergers, scientists can gain insights into the properties of these exotic objects, including their densities, masses, and the nature of their cores.

**Q:** How does gravity influence the behavior of galaxies in the vast cosmic web of the universe?

**A:** Gravity shapes the behavior of galaxies within the cosmic web of the universe. Galaxies are drawn towards one another by gravity, leading to the formation of galaxy clusters and superclusters. The cosmic web's filamentary structure is a direct result of gravity's pull on matter, revealing the large-scale distribution of galaxies.

**Q:** Are there any proposed methods for harnessing gravity as a potential propulsion system for space travel beyond our solar system?

**A:** Various speculative concepts, like advanced propulsion systems, have been proposed for space travel beyond our solar system, but they often involve hypothetical technologies and face significant challenges. While gravity assists can help spacecraft gain speed, alternative approaches like warp drives and wormholes are currently theoretical and require exotic forms of matter or energy not yet discovered.

**Q:** How does the study of gravity contribute to our understanding of the fundamental forces of the universe and the potential unification of these forces?

**A:** The study of gravity contributes to our understanding of the fundamental forces of the universe by highlighting the challenges of unifying gravity with the other three fundamental forces (electromagnetism and the strong and weak nuclear forces). A unified theory of physics, often called the theory of everything, aims to explain these forces within a single framework, deepening our understanding of the cosmos.

**Q:** Can gravity play a role in explaining phenomena in the quantum realm, such as quantum entanglement or wave-particle duality?

**A:** While gravity is extremely weak at the quantum scale, some scientists explore whether it might have subtle effects that are not yet fully understood. However, these potential effects are challenging to detect and remain a subject of ongoing research. Gravity's role in the quantum realm is a key aspect of the quest for a theory of quantum gravity.

Gravity pulls at more than just matter—it pulls at our deepest questions.

**Q:** How does gravity influence the formation and behavior of cosmic phenomena like planetary rings and asteroid belts?

**A:** Gravity influences the formation and behavior of cosmic phenomena like planetary rings and asteroid belts. In the case of rings, gravity helps shape and maintain their structure by keeping ring particles in orbit around their parent planet. For asteroid belts, gravitational interactions can affect the orbits and dynamics of asteroids within the belt.

**Q:** Are there any philosophical or metaphysical aspects to gravity that have intrigued thinkers throughout history?

**A:** Gravity has indeed intrigued philosophers and thinkers throughout history, sparking questions about the nature of space, time, and the interconnectedness of the cosmos. It has been a subject of philosophical contemplation, inspiring ideas about the underlying unity of the universe and the profound relationships between matter, energy, and spacetime.

**Q:** How does gravity affect the behavior of cosmic phenomena like comet tails and the trajectories of space probes?

**A:** Gravity influences the behavior of comet tails and the trajectories of space probes. Comets' tails are affected by the Sun's gravity, which can cause the tails to point away from the Sun. For space probes, gravity assists from celestial bodies are used to alter their trajectories efficiently, allowing them to reach distant destinations with less fuel.

**Q:** Can the study of gravity provide insights into the potential existence of hidden dimensions or alternate universes?

**A:** The study of gravity, particularly in the context of theories like string theory and braneworld cosmology, has raised the possibility of hidden dimensions or alternate universes. These theories propose that gravity might be sensitive to dimensions beyond the familiar three spatial dimensions and one time dimension. However, direct evidence for such dimensions remains elusive.

**Q:** How does the concept of gravitational time dilation, as predicted by general relativity, impact our understanding of time itself?

**A:** Gravitational time dilation, as predicted by general relativity, has profound implications for our understanding of time. It demonstrates that time is not an absolute and uniform concept but can vary depending on the strength of the gravitational field. This effect has been confirmed through experiments, such as the gravitational redshift of light from massive objects, and is a fundamental aspect of modern physics.

**Q:** Can gravity waves provide information about the interior structure and composition of celestial objects, like the Earth's core or the composition of exoplanets?

**A:** Gravitational waves have the potential to provide information about the interior structure and composition of celestial objects. For example, they can reveal details about the distribution of mass within neutron stars and could potentially be used to study the Earth's interior, including its core. However, such studies would require precise gravitational wave measurements and sophisticated analysis techniques.

**Q:** How does the study of gravity contribute to our understanding of the large-scale structure and evolution of the universe, including concepts like cosmic inflation and the Big Bang?

**A:** Gravity is central to our understanding of the large-scale structure and evolution of the universe. It is a key player in cosmic inflation, a theory that explains the rapid expansion of the early universe. Gravity's influence on matter distribution is also crucial for understanding the formation of galaxies and galaxy clusters, leading to insights into the universe's evolution since the Big Bang.

**Q:** Are there any potential practical applications of our understanding of gravity, beyond space exploration and fundamental physics research?

**A:** Our understanding of gravity has practical applications in various fields, such as geophysics, geodesy, and navigation. It is used in technologies like GPS (Global Positioning System) to accurately determine positions on Earth's surface. Additionally, gravitational principles are applied in industries like oil exploration and environmental monitoring.

**Q:** How do theories like quantum gravity and theories of everything seek to reconcile gravity with the fundamental forces of the quantum world?

**A:** Theories like quantum gravity and theories of everything aim to reconcile gravity with the quantum world by providing a consistent framework that unifies all fundamental forces, including gravity. These theories propose new concepts and mathematical formalisms that extend quantum mechanics to incorporate gravity, potentially shedding light on the behavior of gravity at extremely small scales and in extreme environments.

Gravity shapes the universe in profound ways, from the smallest particles to the grandest galaxies.

**Q:** Can gravity have a role in explaining phenomena at the cosmic scale, such as the expansion of the universe and the formation of cosmic structures?

**A:** Gravity plays a central role in explaining phenomena at the cosmic scale. It drives the expansion of the universe, acting as the attractive force that counteracts the universe's expansion due to the initial Big Bang. Gravity's influence on matter also shapes the formation of cosmic structures, from galaxies to galaxy clusters, by drawing matter together over cosmic timescales.

**Q:** How does gravity influence the motion of celestial objects within galaxies, such as stars within our Milky Way?

**A:** Gravity is responsible for the motion of celestial objects within galaxies. In our Milky Way, for example, the gravitational pull of the central supermassive black hole keeps stars in orbit around it. Gravity also determines the orbits of stars around the galactic center and their relative velocities.

**Q:** Can the study of gravity help us understand the behavior of matter and energy in extreme environments, such as the conditions inside neutron stars or during supernova explosions?

**A:** The study of gravity is essential for understanding the behavior of matter and energy in extreme environments. For example, it plays a crucial role in the collapse and explosion of massive stars in supernovae. Gravity also influences the conditions inside dense objects like neutron stars, where immense gravitational forces are at work.

**Q:** How does gravity relate to the concept of spacetime curvature, as described by Einstein's theory of general relativity?

**A:** Gravity is intimately connected to the concept of spacetime curvature in general relativity. According to this theory, massive objects warp or curve the fabric of spacetime around them. Other objects then follow curved paths in response to this curvature, which we perceive as the force of gravity. This idea revolutionized our understanding of gravity, replacing the older concept of gravity as a force acting at a distance.

**Q:** Can the study of gravity reveal insights into the nature of dark matter and dark energy, two mysterious components of the universe?

**A:** The study of gravity is instrumental in our quest to understand dark matter and dark energy. While we cannot directly observe these entities, their gravitational effects are detectable.

Gravity's influence on the motion of galaxies and the expansion of the universe provides crucial evidence for the existence of dark matter and the accelerated expansion driven by dark energy.

Q: How does the concept of escape velocity relate to gravity, and how does it impact space travel and rocket launches?

A: Escape velocity is the minimum velocity required for an object to break free from a celestial body's gravitational pull. It is directly related to the strength of gravity on that body. Understanding escape velocity is essential for space travel; rockets must reach or exceed it to leave Earth's gravitational influence. Different celestial bodies have different escape velocities due to variations in their mass and radius.

Q: Are there any experiments or missions planned to study gravity in novel ways or under extreme conditions, such as in microgravity environments or near massive black holes?

A: Yes, there are ongoing experiments and planned missions to study gravity under novel conditions. For instance, microgravity experiments on the International Space Station (ISS) provide insights into the effects of low gravity on biological and physical systems. Future missions, like the James Webb Space Telescope, will study the gravitational lensing of distant galaxies by massive objects, revealing hidden structures in the universe.

Q: How does gravity influence the behavior of cosmic phenomena like pulsars, which emit beams of radiation as they rotate rapidly?

A: Gravity influences pulsars, which are rapidly rotating neutron stars. Neutron stars are incredibly dense, and their strong gravitational fields can cause the observed behavior of pulsars. Gravity affects the emission of radiation from their magnetic poles, leading to the periodic pulses of radiation that we detect as pulsar signals.

Q: Can gravitational interactions between celestial bodies lead to phenomena like orbital resonances or gravitational captures, where one body's gravity affects the motion of another?

A: Gravitational interactions between celestial bodies can indeed lead to phenomena like orbital resonances and gravitational captures. Orbital resonances occur when two bodies have periodic gravitational interactions that affect their orbits, while gravitational captures happen when one object's gravity captures another into a stable orbit or trajectory.

# In the pull of gravity, we find the freedom to question, wonder, and learn.

**Q:** How does the concept of frame-dragging, also known as the Lense-Thirring effect, illustrate the way massive objects like Earth can "drag" or twist the fabric of spacetime around them?

**A:** Frame-dragging, predicted by general relativity, describes how a massive object, like Earth, can "drag" or twist the fabric of spacetime around it as it rotates. This effect influences the behavior of nearby objects, including the orbits of satellites. It's a manifestation of how gravity and spacetime interact in a dynamic way, revealing the deep connection between mass and the structure of the universe.

**Q:** Can the study of gravity inform our understanding of cosmic phenomena like the formation of planetary systems and the potential habitability of exoplanets?

**A:** Gravity plays a fundamental role in the formation of planetary systems, including our own solar system. The gravitational collapse of a protoplanetary disk leads to the formation of planets and their subsequent orbital dynamics. Understanding these processes informs our search for exoplanets and their potential habitability, as gravity influences factors like planetary size, distance from the host star, and the presence of moons.

**Q:** How does the behavior of gravity near black holes challenge our understanding of physics, particularly the interplay between gravity, quantum mechanics, and information preservation?

**A:** The behavior of gravity near black holes challenges our understanding of physics at its most extreme limits. The study of black hole physics highlights the need for a consistent theory of quantum gravity that can describe the behavior of matter and energy in the intense gravitational fields near black holes. It also raises intriguing questions about the preservation of information, a topic of ongoing research known as the black hole information paradox.

**Q:** Can gravity have implications for the development of advanced propulsion technologies, such as those required for interstellar travel or faster-than-light travel in the future?

**A:** While gravity itself is not a source of propulsion, our understanding of gravity is crucial for the development of advanced propulsion technologies. Concepts like warp drives and wormholes, which are speculative at this point, rely on gravity-related principles. These ideas, inspired by science fiction, are the subject of ongoing research and exploration in theoretical physics.

**Q:** How does the study of gravity contribute to our understanding of the fundamental nature of the universe, including questions about the origin and fate of the cosmos?

**A:** Gravity plays a central role in our understanding of the universe's fundamental nature. It contributes to our knowledge of cosmic evolution, from the early moments after the Big Bang to the ultimate fate of the universe. Questions about cosmic inflation, dark energy, and the cosmic microwave background radiation all involve gravity's influence on the large-scale structure and dynamics of the cosmos.

**Q:** Can gravitational effects explain phenomena like the bending of light by massive objects, leading to gravitational lensing, and how is this phenomenon observed and utilized in astronomy?

**A:** Gravitational lensing, a phenomenon predicted by general relativity, occurs when the gravitational field of a massive object, such as a galaxy or a cluster of galaxies, bends the path of light from a distant source. This bending effect can create multiple images or distortions of the source. Gravitational lensing is observed in astronomy and is used to study distant objects, including the discovery of exoplanets and the mapping of dark matter distribution.

**Q:** How does the study of gravity contribute to the exploration of the early universe, including the conditions shortly after the Big Bang and the formation of the first cosmic structures?

**A:** Gravity plays a pivotal role in understanding the early universe. It is instrumental in modeling the conditions shortly after the Big Bang, including the expansion of the universe and the formation of primordial structures. By studying the cosmic microwave background radiation and large-scale cosmic structures, scientists gain insights into the universe's early history, which is largely shaped by gravity's influence.

**Q:** How does the concept of gravitational time dilation affect the way we perceive the age of distant cosmic objects and events, such as light from distant galaxies or cosmic microwave background radiation?

**A:** Gravitational time dilation, as predicted by general relativity, means that time passes differently in regions with different gravitational strengths. This effect can lead to differences in the perceived ages of cosmic objects. For example, light from distant galaxies may appear older due to the time dilation experienced as it travels through space and encounters various gravitational fields.

**Q:** Can the study of gravitational waves provide insights into the properties and behaviors of neutron stars, including their extreme densities and exotic states of matter?

**A:** Yes, the study of gravitational waves can provide insights into neutron stars, which are among the densest objects in the universe. When neutron stars merge, they emit gravitational waves that carry information about their masses, spins, and the properties of matter at extreme densities. Analyzing these waves can help us better understand the nature of neutron stars and the behavior of matter under such extreme conditions.

Gravity: bending
light, shaping stars, and
inspiring minds.

**Q:** How does the phenomenon of gravitational lensing, where massive objects bend light, assist astronomers in discovering and studying distant galaxies, dark matter, and even planets outside our solar system?

**A:** Gravitational lensing is a powerful tool in astronomy. When massive objects act as gravitational lenses, they bend and magnify light from objects behind them. This effect can be used to discover distant galaxies that would otherwise be too faint to detect. Gravitational lensing can also reveal the distribution of dark matter within galaxy clusters and has been employed in the discovery of exoplanets through a technique known as microlensing.

**Q:** Can gravity assist in our understanding of the fundamental particles and forces of the universe, as explored in the field of particle physics and high-energy experiments?

**A:** Gravity is a weak force compared to the other fundamental forces, such as electromagnetism and the strong and weak nuclear forces. While gravity is not a direct focus of particle physics experiments, its influence is accounted for in particle interactions. However, at extremely high energies, such as those near the Planck scale, the interplay between gravity and particle physics becomes crucial, leading to the quest for a theory of quantum gravity.

**Q:** How does gravity impact the behavior of cosmic phenomena like quasars, which are incredibly energetic and luminous objects at the centers of distant galaxies?

**A:** Gravity plays a significant role in the behavior of quasars. Quasars are powered by the infall of matter onto supermassive black holes at the centers of galaxies. The gravitational potential energy released in this process fuels the immense luminosity of quasars. Studying quasars provides insights into the dynamics of matter around black holes and the strong gravitational fields present in these regions.

**Q:** Can the study of gravity contribute to our understanding of the ultimate fate of the universe, including scenarios like the Big Crunch or the eternal expansion driven by dark energy?

**A:** Gravity is central to our understanding of the universe's fate. It influences the expansion rate of the universe, which is determined by the interplay between the attractive force of gravity and the repulsive force of dark energy. Depending on the balance between these forces, the universe could experience scenarios like the Big Crunch, eternal expansion, or a stable state.

**Q:** How does gravity impact the behavior of matter and radiation in the cosmos, leading to the formation of structures like galaxies, galaxy clusters, and cosmic filaments?

**A:** Gravity is the driving force behind the formation of structures in the universe. Over cosmic timescales, gravity causes matter to clump together, forming structures from small galaxies to massive galaxy clusters and even the large-scale cosmic web of filaments. The study of these structures provides insights into the distribution of matter and the history of the universe's evolution.

**Q:** Can gravity have implications for our understanding of the fundamental constants of nature, such as the gravitational constant (G), and their potential variations across the universe?

**A:** Gravity and the fundamental constants of nature are interconnected. The gravitational constant (G) is one of these constants, and its value affects the strength of gravity. Variations in fundamental constants are a subject of interest in cosmology and fundamental physics. The study of these variations can shed light on the universe's history and the underlying laws of physics.

**Q:** How does gravity influence the formation and evolution of cosmic phenomena like star clusters, globular clusters, and galaxies?

**A:** Gravity plays a central role in the formation and evolution of these cosmic phenomena. Star clusters form as gravity causes groups of stars to come together. Globular clusters are dense collections of stars bound by gravity. On a larger scale, galaxies are massive systems of stars, gas, and dust held together by gravity. The study of their dynamics and interactions provides insights into their evolution.

**Q:** Can gravity be harnessed for practical applications in space, such as propelling spacecraft, generating power, or assisting in resource extraction from asteroids or other celestial bodies?

**A:** Gravity can be harnessed for various practical applications in space. Gravity assists from planets are used to propel spacecraft on interplanetary missions. Gravitational gradients can be used to generate power through techniques like tidal power. In the future, gravity could play a role in resource extraction from celestial bodies, such as mining asteroids.

Knowing gravity isn't aabout force—it's a force that shapes our questions.

**Q:** How does the concept of event horizons, as associated with black holes, challenge our understanding of space and time and raise intriguing questions about the fate of matter and information within them?

**A:** Event horizons, the boundaries beyond which nothing can escape a black hole's gravitational pull, challenge our understanding of space and time. They create a point of no return, where the laws of physics as we know them break down. The fate of matter and information that enters black holes is a subject of intense theoretical research, leading to questions about whether information is lost or preserved.

**Q:** Can gravity assist us in exploring the interiors of planets, such as Mars or Jupiter, by measuring gravitational variations that reveal details about their internal structures and compositions?

**A:** Yes, gravity can assist in exploring the interiors of planets. By measuring gravitational variations as spacecraft orbit these planets, scientists can infer information about their internal structures, including the distribution of mass and density. This technique has been used in missions to Mars, Earth, and other celestial bodies to gain insights into their subsurface properties.

**Q:** How does gravity impact the behavior of cosmic phenomena like cosmic microwave background radiation, which provides a glimpse into the early universe's conditions and evolution?

**A:** Gravity influences the behavior of cosmic microwave background radiation (CMB), which is the afterglow of the Big Bang. Gravitational interactions cause slight temperature fluctuations in the CMB, which reveal details about the distribution of matter in the early universe. Studying these fluctuations helps us understand the universe's early conditions and the seeds of cosmic structures.

**Q:** Can the study of gravity provide insights into the potential existence of exotic cosmic objects like primordial black holes or cosmic strings, which are theorized but remain undetected?

**A:** The study of gravity can indeed provide insights into the potential existence of exotic cosmic objects like primordial black holes and cosmic strings. Researchers search for gravitational signatures and effects these objects may produce, such as gravitational lensing or gravitational waves. Detecting or ruling out their existence would have significant implications for our understanding of the universe.

**Q:** How does gravity contribute to the behavior of cosmic phenomena like gamma-ray bursts and supernovae, which release immense amounts of energy and influence the chemical composition of the cosmos?

**A:** Gravity is intimately linked to the behavior of cosmic phenomena like gamma-ray bursts and supernovae. Massive stars undergo gravitational collapse before exploding as supernovae, releasing enormous energy. Gamma-ray bursts often result from these explosions. These events not only shape the chemical composition of the cosmos by synthesizing elements but also serve as cosmic beacons that can be observed across vast cosmic distances.

**Q:** Can gravity have implications for understanding the nature of dark energy, the mysterious force responsible for the universe's accelerated expansion, and its potential connection with vacuum energy or a cosmological constant?

**A:** Gravity is essential for understanding dark energy, which is associated with the universe's accelerated expansion. The behavior of gravity, particularly its interplay with dark energy, is central to cosmological models. Dark energy may be linked to vacuum energy or a cosmological constant, and its study sheds light on the fate and geometry of the universe.

**Q:** How does gravity influence the formation and behavior of planetary rings, like those around Saturn, and what can studying these rings reveal about the dynamics of celestial systems?

**A:** Gravity plays a key role in the formation and behavior of planetary rings. These rings consist of countless particles held in orbit around a planet by gravity. Studying these rings can reveal insights into the dynamics of celestial systems, including the interactions between individual ring particles and the processes that shape the rings over time.

**Q:** Can the study of gravity help us understand the behavior of cosmic phenomena like supernova remnants and the distribution of heavy elements in galaxies, which are essential for the formation of planets and life as we know it?

**A:** Yes, the study of gravity is integral to understanding phenomena like supernova remnants and the distribution of heavy elements in galaxies. Supernovae, powered by gravitational collapse, are responsible for producing and dispersing heavy elements into space. These elements then play a crucial role in the formation of planets and life. By studying these processes, scientists gain insights into the cosmic origins of the elements essential for life.

The force that shapes stars,
bends time, and fuels our
quest to understand the
universe.

**Q:** How does gravity affect the behavior of cosmic dust and gas clouds in space, leading to the formation of stars and planetary systems, and what can this tell us about the prevalence of habitable environments in the universe?

**A:** Gravity is the driving force behind the collapse of cosmic dust and gas clouds, leading to the formation of stars and planetary systems. Understanding this process is vital for assessing the prevalence of habitable environments in the universe. The study of star formation and planetary system formation helps us explore the conditions necessary for life to emerge on planets within these systems.

**Q:** Can gravity assist in the search for extraterrestrial life by helping scientists identify potentially habitable exoplanets based on their mass, composition, and orbital characteristics?

**A:** Gravity is a crucial factor in the search for extraterrestrial life. By studying exoplanets' mass, composition, and orbital characteristics, scientists can assess their potential habitability. Gravity affects factors like surface pressure, temperature, and the retention of atmospheres, all of which are essential considerations when evaluating a planet's suitability for life.

**Q:** How does gravity impact the behavior of matter in extreme environments like quark-gluon plasmas, which are thought to have existed shortly after the Big Bang, and what can the study of these environments reveal about the early universe?

**A:** Gravity's role in extreme environments like quark-gluon plasmas is a subject of great interest in understanding the early universe. Such plasmas are believed to have existed shortly after the Big Bang when conditions were extremely hot and dense. Gravity influences the expansion and behavior of matter in these environments, and studying them can provide insights into the universe's earliest moments and the transition from quark-gluon plasma to the formation of matter as we know it.

**Q:** Can gravity assist in predicting and preparing for potential cosmic hazards, such as asteroid impacts or gravitational interactions with nearby celestial bodies that could affect Earth?

**A:** Gravity plays a role in predicting potential cosmic hazards. By tracking the orbits of asteroids and other celestial bodies, scientists can use gravitational calculations to predict their future paths and assess the risk of impacts. This information is crucial for planetary defense and preparedness efforts aimed at mitigating the potential impact of such hazards.

**Q:** How does gravity influence the behavior of matter in the interstellar medium, where stars are born and die, and what can this tell us about the life cycles of stars and galaxies?

**A:** Gravity is instrumental in shaping the behavior of matter in the interstellar medium, where stars are born and where stellar remnants contribute to the cycle of matter in galaxies. Gravity governs the collapse of molecular clouds to form new stars and plays a role in the dynamics of galaxies. Studying the interstellar medium provides valuable insights into the life cycles of stars and the evolution of galaxies over cosmic timescales.

**Q:** Can gravity be harnessed for future space exploration and colonization efforts, such as creating artificial gravity in spacecraft or establishing sustainable habitats on celestial bodies like the Moon or Mars?

**A:** Gravity can be harnessed for future space exploration and colonization efforts. Creating artificial gravity in spacecraft or space stations through centrifugal force is one approach to mitigating the health effects of prolonged space travel. Additionally, understanding gravity's influence on celestial bodies like the Moon or Mars is essential for planning sustainable habitats and resource utilization in these environments.

**Q:** How does gravity impact the behavior of cosmic rays, which are high-energy particles from space, and what can the study of cosmic rays reveal about the universe's most extreme environments?

**A:** Gravity influences the trajectories of cosmic rays as they travel through the cosmos. Cosmic rays provide a unique window into extreme environments like supernova remnants, pulsars, and even black holes. By studying cosmic rays, scientists gain insights into the most energetic and extreme processes in the universe.

**Q:** Can gravity have implications for the stability and behavior of planetary atmospheres, including their role in regulating temperature, weather patterns, and potential habitability on Earth and other celestial bodies?

**A:** Gravity plays a crucial role in the stability and behavior of planetary atmospheres. It determines a planet's ability to retain its atmosphere, which, in turn, impacts its climate and potential for habitability. On Earth, gravity is responsible for maintaining the atmosphere and stabilizing the climate, making it essential for supporting life as we know it.

**Q:** How does gravity contribute to the understanding of the expansion of the universe, including the measurement of the Hubble constant and the rate at which the cosmos is expanding, and what do these measurements reveal about the universe's age and fate?

**A:** Gravity is central to our understanding of the expansion of the universe. The rate of this expansion, described by the Hubble constant, is influenced by the interplay between gravity's attractive force and the repulsive force of dark energy. Measurements of the Hubble constant provide crucial information about the universe's age and the balance between these cosmic forces, shedding light on its eventual fate.

**Gravity lets us observe and study the universe in reolutionary ways, i.e. futuristic Gravity / Space-Time Based Computers**

**Q:** Can gravity assist in the detection and study of exoplanets, including their potential atmospheres and habitability, through techniques like gravitational microlensing or transit observations?

**A:** Yes, gravity plays a role in the detection and study of exoplanets. Gravitational microlensing can reveal the presence of exoplanets through the bending of light by their gravity. Transit observations, which involve the periodic dimming of a star's light as an exoplanet passes in front of it, provide information about exoplanet atmospheres and potential habitability.

**Q:** How does the concept of gravitational waves revolutionize our ability to observe and study the universe, including the merger of black holes and neutron stars, and what can these observations reveal about the extreme events shaping the cosmos?

**A:** Gravitational waves have revolutionized our ability to observe the universe. They allow us to detect and study extreme events like the mergers of black holes and neutron stars. These observations provide direct evidence of gravitational interactions in the most extreme environments, offering insights into the behavior of matter and gravity under such conditions.

**Q:** Can gravity have implications for the behavior of light and electromagnetic radiation in the cosmos, leading to phenomena like gravitational lensing, which distorts and magnifies the appearance of distant objects, and how is this phenomenon utilized in astronomy?

**A:** Gravity's influence on light and electromagnetic radiation is evident in phenomena like gravitational lensing, which has been used in astronomy to study distant objects. Gravitational lensing can magnify the appearance of background galaxies, allowing astronomers to observe details that would otherwise be too faint to detect. This technique has contributed to our understanding of dark matter distribution and the discovery of exoplanets.

**Q:** How does gravity impact the behavior of cosmic structures like galaxy clusters, superclusters, and voids in the large-scale structure of the universe, and what can their study reveal about the cosmic web and the distribution of matter on the largest scales?

**A:** Gravity governs the behavior of cosmic structures on the largest scales. It leads to the formation of galaxy clusters, superclusters, and voids within the cosmic web. Studying the distribution of matter and the dynamics of these structures provides insights into the large-scale structure of the universe, the nature of dark matter, and the evolution of cosmic filaments over billions of years.

**Q:** Can gravity be used to probe the properties of fundamental particles, such as neutrinos, by studying their interactions with cosmic phenomena like supernovae or by conducting precision experiments on Earth?

**A:** Gravity can be used to probe the properties of fundamental particles like neutrinos. Neutrinos interact very weakly with matter, but their mass and properties influence their behavior in gravitational fields. Studying neutrinos emitted from cosmic phenomena like supernovae or conducting precision experiments on Earth allows scientists to gain insights into these elusive particles and their role in the universe.

**Q:** How does gravity affect the formation and stability of planetary orbits, and what can the study of orbital dynamics reveal about the long-term stability of our solar system and others in the galaxy?

**A:** Gravity is the primary force shaping planetary orbits. Studying orbital dynamics provides insights into the long-term stability of our solar system and others. It can reveal potential interactions and resonances that influence the orbits of planets and other celestial bodies over billions of years.

**Q:** How does gravity influence the behavior of cosmic jets produced by objects like active galactic nuclei (AGN) and quasars, and what can the study of these powerful phenomena tell us about the energy release mechanisms near supermassive black holes?

**A:** Gravity plays a role in shaping the behavior of cosmic jets produced by AGN and quasars. These high-energy phenomena involve the interplay of gravity, magnetic fields, and relativistic particles near supermassive black holes. Studying cosmic jets provides insights into the extreme conditions and energy release mechanisms near these massive objects.

**Q:** Can gravity help us understand the nature of dark matter, an elusive substance that does not emit light or energy, through its gravitational effects on the visible matter in galaxies and galaxy clusters?

**A:** Yes, gravity plays a crucial role in our quest to understand dark matter. Dark matter's gravitational effects on visible matter, such as galaxies and galaxy clusters, are a primary means of detection. By studying these gravitational interactions, scientists can map the distribution of dark matter in the universe and infer its properties.

**Q:** How does gravity contribute to the study of cosmic microwave background radiation (CMB), the relic radiation from the early universe, and what can it reveal about the universe's initial conditions and evolution?

## Exploring how gravity shapes innovation

**A:** Gravity influences the behavior of matter and radiation in the early universe, which is reflected in the CMB. Variations in temperature in the CMB provide a snapshot of the universe's initial conditions and evolution. Studying the CMB allows scientists to probe the universe's early moments and the seeds of cosmic structures.

**Q:** Can gravity be used to investigate the existence of additional dimensions beyond the familiar three spatial dimensions and one time dimension, as hypothesized in theories like string theory or brane cosmology?

**A:** Yes, gravity can be used to explore the existence of additional dimensions. Some theories, such as string theory, propose the existence of extra dimensions beyond our familiar four. Gravitational interactions in these theories can manifest differently from standard gravity, providing potential experimental clues to the existence of extra dimensions.

**Q:** How does gravity influence the behavior of matter and energy in the vicinity of neutron stars, which are incredibly dense remnants of massive stellar explosions, and what can studying these objects reveal about the fundamental properties of matter under extreme conditions?

**A:** Gravity has a profound impact on matter and energy near neutron stars. These objects are incredibly dense, and their intense gravitational fields can warp spacetime. Studying neutron stars provides insights into the behavior of matter at extreme densities and the properties of the densest states of matter in the universe.

**Q:** How does gravity contribute to the formation and dynamics of galaxy clusters, the largest gravitationally bound structures in the universe, and what can their study tell us about the distribution of matter on cosmic scales and the influence of dark matter?

**A:** Gravity is the driving force behind the formation and dynamics of galaxy clusters. These clusters are the largest gravitationally bound structures in the universe. Studying them provides valuable information about the distribution of matter on cosmic scales and the role of dark matter in the cosmic web's formation and evolution.

**Q:** Can gravity assist in understanding the behavior of cosmic strings, hypothetical one-dimensional topological defects in the fabric of spacetime, and what implications could the detection or study of cosmic strings have for our understanding of fundamental physics and the early universe?

**A:** Gravity can play a role in understanding the behavior of cosmic strings, should they exist. Cosmic strings are hypothetical objects that could form in the early universe. Their gravitational effects, such as gravitational lensing, may provide indirect clues to their existence. The study of cosmic strings could have profound implications for our understanding of fundamental physics and the early universe's conditions.

**Q:** How does gravity influence the behavior of binary star systems, where two stars orbit around their common center of mass, and what can these systems tell us about the masses, lifecycles, and dynamics of stars?

**A:** Gravity governs the dynamics of binary star systems, offering insights into the masses, lifecycles, and interactions of stars. By studying binary stars, astronomers can measure stellar masses with great precision and uncover details about stellar evolution and the effects of gravitational interactions.

**Q:** Can gravity play a role in the search for evidence of extraterrestrial life, such as the study of subsurface oceans on icy moons like Europa or Enceladus, where gravitational interactions with their parent planets generate heat and maintain liquid water?

**A:** Yes, gravity is a key factor in the search for extraterrestrial life. On moons like Europa and Enceladus, gravitational interactions with their parent planets generate tidal heating, maintaining subsurface oceans. These oceans could potentially harbor life, making the study of gravitational effects vital in assessing habitability.

**Q:** How does gravity affect the behavior of cosmic dust grains and particles in the interstellar medium, influencing processes like star and planet formation, and what can this reveal about the origins of celestial bodies in the universe?

**A:** Gravity shapes the behavior of cosmic dust and particles, enabling processes like star and planet formation. Understanding these interactions helps astronomers trace the origins of celestial bodies and the mechanisms by which they come into existence.

Harnessing the elegance of gravity's
silent dance to propel us into
the unknown.

**Q:** Can gravity provide insights into the mysterious nature of dark energy, including its role in the accelerated expansion of the universe, and the potential interplay between dark energy, gravity, and cosmic structures?

**A:** Gravity is integral to understanding dark energy's role in the accelerated expansion of the universe. The interaction between gravity and dark energy is a central focus of cosmological research, shedding light on the universe's fate and the formation of cosmic structures.

**Q:** How does gravity influence the dynamics of galaxies within galaxy clusters, leading to phenomena like galactic cannibalism and the formation of massive central galaxies, and what can this tell us about the hierarchical growth of cosmic structures?

**A:** Gravity plays a significant role in the dynamics of galaxies within galaxy clusters, leading to processes like galactic cannibalism and the formation of massive central galaxies. Studying these interactions provides insights into the hierarchical growth of cosmic structures and the evolution of galaxy clusters.

**Q:** Can gravity assist in the detection and study of cosmic phenomena like gamma-ray bursts, which release intense bursts of high-energy radiation, and the potential role of neutron stars or black holes in their origins?

**A:** Yes, gravity is involved in the study of cosmic phenomena like gamma-ray bursts. These events are thought to result from processes involving neutron stars or black holes. Understanding the gravitational effects in these extreme environments helps astronomers unravel the origins of gamma-ray bursts.

**Q:** How does gravity impact the behavior of exotic cosmic objects like white dwarfs, which are the remnants of lower-mass stars, and what can studying these objects reveal about the endpoints of stellar evolution and the fate of our Sun?

**A:** Gravity influences the behavior of exotic cosmic objects like white dwarfs, which represent the endpoints of lower-mass stars. Studying these objects provides insights into the final stages of stellar evolution and offers a glimpse into the future fate of our Sun.

**Q:** Can gravity assist in the investigation of cosmic phenomena like planetary migration, where planets within a solar system change their orbits over time, and what implications does this have for our understanding of planetary system formation and stability?

**A:** Gravity is a central factor in the study of planetary migration, which can significantly impact planetary system formation and stability. Understanding the gravitational interactions within planetary systems helps unravel the mechanisms behind planetary migration and its consequences.

**Q:** How does gravity contribute to the study of exotic cosmic phenomena like pulsars and magnetars, which are highly magnetized neutron stars with extreme gravitational fields, and what can these objects reveal about the behavior of matter under extreme conditions?

**A:** Gravity is a critical component of the study of exotic cosmic objects like pulsars and magnetars. These objects have intense gravitational fields that warp spacetime. Studying them provides insights into the behavior of matter under the most extreme conditions in the universe.

**Q:** How does gravity influence the behavior of cosmic phenomena like planetary rings, where particles orbit a central planet, and what can the study of ring systems reveal about the formation and stability of planetary systems?

**A:** Gravity plays a crucial role in the behavior of planetary rings, where particles are held in orbit around a central planet. Studying these rings provides insights into the formation and stability of planetary systems, including the interplay between gravity and other forces that shape these structures.

**Q:** Can gravity assist in our understanding of the fundamental forces of the universe, including the unification of gravity with the other fundamental forces in theories like quantum gravity, and what implications could such a unification have for our understanding of the early universe and particle physics?

**A:** Gravity is central to the quest for a unified theory that combines it with the other fundamental forces. The unification of gravity with the strong, weak, and electromagnetic forces is a longstanding goal in physics, potentially revealing connections between the early universe and particle physics. Such a theory, like a theory of quantum gravity, would have profound implications for our understanding of the fundamental forces at play in the cosmos.

# Gravitational wave communication

**Q:** How does gravity impact the behavior of cosmic phenomena like the Kuiper Belt, a region beyond Neptune containing small icy bodies, and what can the study of these objects reveal about the formation and evolution of our solar system and the potential for hidden planets or dwarf planets?

**A:** Gravity shapes the behavior of objects in the Kuiper Belt, influencing their orbits and interactions. Studying these objects provides insights into the formation and evolution of our solar system. It can also lead to the discovery of hidden planets or dwarf planets that may exist in the distant reaches of the solar system.

**Q:** Can gravity help us understand the existence of stable Lagrange points in celestial systems, where gravitational forces balance and allow objects to remain in fixed positions relative to two massive bodies, and what applications do these points have in space exploration and satellite positioning?

**A:** Gravity is instrumental in the existence of stable Lagrange points in celestial systems. These points are crucial in space exploration and satellite positioning, allowing spacecraft to maintain fixed positions relative to massive bodies like planets. Applications include space observatories, communication satellites, and space missions that leverage these stable points for various purposes.

**Q:** How does gravity influence the behavior of cosmic dust and gas clouds in the interstellar medium, leading to the formation of new stars and the replenishment of cosmic material, and what can this tell us about the continuous cycle of stellar birth and death in the galaxy?

**A:** Gravity plays a pivotal role in the behavior of cosmic dust and gas clouds in the interstellar medium, leading to the formation of new stars. Understanding this process offers insights into the continuous cycle of stellar birth and death in the galaxy, as well as the replenishment of cosmic material necessary for star and planet formation.

**Q:** Can gravity assist in the study of planetary magnetic fields, including their generation, interactions with cosmic radiation, and effects on planetary habitability, by revealing information about the internal structure and composition of planets like Earth and Jupiter?

**A:** Yes, gravity can contribute to the study of planetary magnetic fields. The internal structure and composition of planets influence their gravitational fields, providing clues about the generation and behavior of magnetic fields. Understanding planetary magnetic fields is essential for assessing habitability and protecting against cosmic radiation.

**Q:** How does gravity impact the behavior of matter and radiation near black holes, where intense gravitational forces create extreme conditions, and what can the study of these regions reveal about the nature of spacetime and the boundaries of our understanding of physics?

**A:** Gravity near black holes creates extreme conditions, affecting the behavior of matter and radiation. Studying these regions, including the event horizon, offers insights into the nature of spacetime, the behavior of matter under extreme gravity, and the boundaries of our current understanding of physics, including the laws of thermodynamics and quantum mechanics.

**Q:** Can gravity be used to probe the interior structures of planets and moons, including their core composition and seismic activity, by studying the gravitational variations at their surfaces, and what can this reveal about the geological history of these celestial bodies?

**A:** Gravity is a valuable tool for probing the interior structures of planets and moons. Gravitational variations at their surfaces provide information about core composition, seismic activity, and geological history. This technique has been used in planetary science to uncover details about the internal characteristics of celestial bodies in our solar system.

**Q:** How does gravity influence the behavior of cosmic phenomena like the Oort Cloud, a hypothetical region of comets surrounding the solar system, and what can the study of comets from the Oort Cloud reveal about the history and composition of our solar system?

**A:** Gravity plays a role in the dynamics of comets within the Oort Cloud. Studying comets originating from this region can provide insights into the history and composition of our solar system, including the conditions prevailing during its formation.

**Q:** Can gravity assist in the study of cosmic structures like supermassive black holes at the centers of galaxies, and how do gravitational interactions within these structures contribute to phenomena like active galactic nuclei (AGN) and quasars?

**A:** Gravity is central to the study of supermassive black holes at the centers of galaxies. Gravitational interactions within these structures drive processes like the formation of accretion disks and the generation of intense radiation in AGN and quasars. Understanding these interactions sheds light on the behavior of matter under extreme gravitational conditions.

**Q:** How does gravity influence the behavior of cosmic dust and gas in protoplanetary disks, where planetary systems are born, and what can this tell us about the conditions necessary for the formation of habitable planets and the diversity of planetary systems in the universe?

**A:** Gravity shapes the behavior of cosmic dust and gas in protoplanetary disks, facilitating the formation of planets and planetary systems. The study of these disks provides insights into the conditions required for the birth of habitable planets and the variety of planetary systems that can emerge in the universe.

**Q:** Can gravity be used to explore the existence and properties of exotic cosmic phenomena like cosmic strings, which are hypothetical one-dimensional topological defects in spacetime, and how could their discovery impact our understanding of the early universe and the fabric of spacetime itself?

**A:** Gravity may play a role in the study of cosmic strings, should they exist. These hypothetical one-dimensional defects could affect spacetime in unique ways. Detecting or studying cosmic strings could have profound implications for our understanding of the early universe and the nature of spacetime.

# ARTIFICIAL GRAVITY
## VIA SPACETIME MANIPULATION

**Q:** How does gravity impact the behavior of cosmic dust and gas in the vicinity of red giants and white dwarfs, two stages in the lifecycle of stars, and what can the study of these regions reveal about the fate of our Sun and the recycling of cosmic material between stellar generations?

**A:** Gravity influences the behavior of cosmic dust and gas around red giants and white dwarfs, representing different stages in stellar evolution. Studying these regions provides insights into the fate of our Sun and the recycling of cosmic material between stellar generations, contributing to our understanding of the cosmic ecosystem.

**Q:** Can gravity assist in the investigation of the gravitational lensing effects produced by galaxy clusters, where the immense gravitational fields magnify and distort the appearance of background galaxies, and how can this phenomenon be used to study the distribution of dark matter within clusters and the expansion rate of the universe?

**A:** Gravity is fundamental to the study of gravitational lensing by galaxy clusters. This phenomenon offers a powerful tool for studying the distribution of dark matter within clusters, as well as measuring the expansion rate of the universe through cosmic lensing effects. It enables astronomers to map the unseen dark matter structures within clusters.

**Q:** How does gravity impact the dynamics of cosmic phenomena like supernovae, which are explosive events marking the end of massive stars, and what can the study of supernovae reveal about the elements they produce, their role in seeding galaxies with heavy elements, and the cosmic cycle of stellar birth and death?

**A:** Gravity influences the dynamics of supernovae, which play a crucial role in seeding galaxies with heavy elements. Studying these explosive events provides insights into the elements they produce, their contribution to the enrichment of cosmic material, and the ongoing cycle of stellar birth and death that shapes the universe.

**Q:** Can gravity be harnessed for future space exploration and propulsion technologies, such as gravitational assists from planets or solar sails that utilize the pressure of sunlight, to enable ambitious missions to explore the outer reaches of our solar system and beyond?

**A:** Gravity assists from planets and innovative technologies like solar sails harness gravity for future space exploration. These techniques enable ambitious missions to explore distant regions of our solar system and venture beyond it. Utilizing gravity as a resource in space travel continues to open up exciting possibilities for scientific exploration and interstellar missions.

**Q:** How does gravity influence the behavior of cosmic phenomena like galactic mergers, where two or more galaxies collide and eventually merge into a single entity, and what can the study of these spectacular events reveal about galaxy evolution and the distribution of dark matter within galaxies?

**A:** Gravity is the driving force behind galactic mergers, shaping the interactions and eventual fusion of galaxies. Studying these cosmic collisions provides insights into galaxy evolution, the formation of structures in the universe, and the role of dark matter in influencing these processes.

**Q:** Can gravity be utilized to investigate the behavior of cosmic strings, hypothetical one-dimensional defects in spacetime, through the study of their gravitational lensing effects on background objects, and what implications could the detection of cosmic strings have for our understanding of fundamental physics and the early universe?

**A:** Gravity can indeed be used to study cosmic strings through their gravitational lensing effects on background objects. Detecting cosmic strings would have profound implications for our understanding of fundamental physics and the early universe, potentially revealing insights into the fabric of spacetime itself and the conditions prevailing during the universe's infancy.

**Q:** How does gravity impact the behavior of cosmic phenomena like the Roche limit, a critical distance at which tidal forces from a massive body can disrupt a smaller object, such as a moon or a comet, and what can this phenomenon tell us about the dynamics of celestial bodies within planetary systems?

**A:** Gravity is responsible for the Roche limit, which defines the critical distance at which tidal forces can disrupt smaller celestial bodies. Understanding this phenomenon offers insights into the dynamics of moons, comets, and other objects within planetary systems. It can also help explain the formation and eventual fate of planetary rings.

**Q:** Can gravity assist in the study of the cosmic microwave background radiation (CMB) and its anisotropies, which provide a snapshot of the early universe, and what information can these fluctuations reveal about the initial conditions and evolution of the universe shortly after the Big Bang?

**A:** Gravity plays a role in the study of the CMB and its anisotropies, which offer a glimpse into the early universe. Variations in the CMB temperature and anisotropies provide valuable information about the initial conditions and evolution of the universe shortly after the Big Bang, including the distribution of matter and energy during that epoch.

Harnessing the power of gravity
to extract energy from black holes

**Q:** How does gravity influence the behavior of cosmic phenomena like tidal forces, which can induce deformations in celestial bodies, such as stretching and squeezing of moons and planets, and what can the study of tidal forces reveal about the internal structures and geological activity of these objects?

**A:** Gravity is responsible for tidal forces, which induce deformations in celestial bodies. Studying tidal forces provides insights into the internal structures and geological activity of moons and planets. These forces can lead to phenomena like tidal heating, which can drive geological processes and influence the evolution of planetary bodies.


**Q:** Can gravity assist in the investigation of the role of cosmic magnetic fields in shaping the behavior of charged particles in space, including their interactions with cosmic radiation and the formation of astrophysical jets and plasmas?

**A:** Gravity does play a role in studying cosmic magnetic fields and their effects on charged particles in space. These interactions are crucial for understanding astrophysical phenomena like the formation of powerful jets, the behavior of cosmic radiation, and the dynamics of plasmas in space environments.


**Q:** How does gravity influence the behavior of cosmic phenomena like gravitational instabilities in accretion disks, which can lead to the formation of stars or planets, and what can the study of these instabilities reveal about the processes of celestial body formation and the distribution of matter in the universe?

**A:** Gravity is a key factor in the occurrence of gravitational instabilities in accretion disks. Studying these instabilities provides insights into the processes of star and planet formation, as well as the distribution of matter in the universe. These phenomena are central to understanding the birth of celestial bodies.


**Q:** Can gravity assist in the exploration of the properties of exotic cosmic phenomena like primordial black holes, which are hypothetical black holes formed in the early universe, and how might their detection impact our understanding of dark matter, the nature of cosmic microwave background anomalies, and the origins of structure in the universe?

**A:** Gravity may play a role in exploring primordial black holes, should they exist. Their detection could have significant implications for our understanding of dark matter, the origins of structure in the universe, and the resolution of cosmic microwave background anomalies. Studying these exotic objects may provide insights into the universe's early moments.

**Q:** How does gravity influence the behavior of cosmic phenomena like the Great Attractor, a mysterious gravitational anomaly that affects the motion of galaxies in our local supercluster, and what can the study of this phenomenon reveal about the large-scale structure and dynamics of the universe?

**A:** Gravity is central to understanding the Great Attractor's influence on galaxy motion in our local supercluster. Studying this phenomenon provides insights into the large-scale structure and dynamics of the universe, offering clues about the distribution of matter on cosmic scales.

**Q:** Can gravity assist in the study of cosmic filaments, the vast, thread-like structures that connect galaxy clusters in the cosmic web, and how does the gravitational interaction within these filaments contribute to the flow of matter in the universe and the formation of cosmic structures?

**A:** Gravity plays a significant role in the study of cosmic filaments, which serve as channels for the flow of matter in the cosmic web. Understanding the gravitational interactions within these filaments helps elucidate the processes driving the formation of cosmic structures, from galaxy clusters to voids.

**Q:** How does gravity influence the behavior of cosmic phenomena like cosmic microwave background (CMB) anomalies, such as the Cold Spot, a large, unusually cold region in the CMB, and what can the study of these anomalies reveal about the early universe, the nature of dark matter, and the cosmic inflation theory?

**A:** Gravity affects the distribution of matter and energy in the early universe, potentially leading to CMB anomalies like the Cold Spot. Studying these anomalies provides insights into the early universe's conditions, the role of dark matter, and the validity of cosmic inflation theory in explaining the universe's rapid expansion.

**Q:** Can gravity be used to explore the nature of cosmic voids, vast regions of the universe with relatively low matter density, and how do gravitational interactions within these voids influence the cosmic web's overall structure and the motion of galaxies within it?

**A:** Gravity plays a role in the formation and dynamics of cosmic voids. Gravitational interactions within voids influence the cosmic web's structure and the motion of galaxies within it. Studying these voids provides valuable information about the large-scale distribution of matter in the universe.

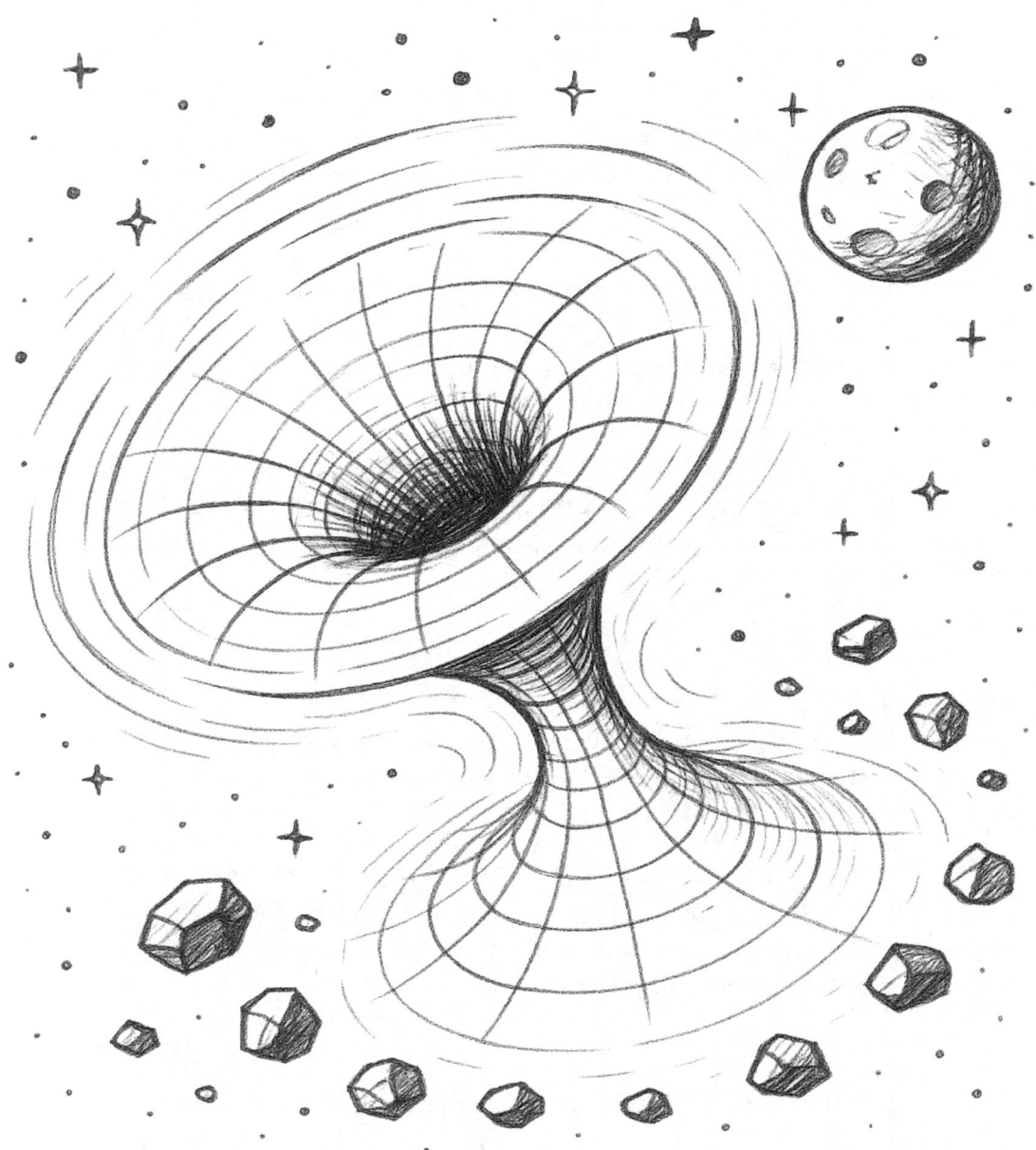

The concept of traversaable
wormholes fires the
imagination.

**Q:** How does gravity impact the behavior of cosmic phenomena like gamma-ray bursts, which release intense bursts of high-energy radiation from distant regions of the universe, and how can the study of these phenomena reveal details about the most energetic events in the cosmos and the properties of their sources?

**A:** Gravity is involved in the study of gamma-ray bursts, which result from highly energetic events in the universe. Understanding gravitational effects in these regions provides insights into the nature of these events, their sources, and the extreme conditions that produce intense bursts of high-energy radiation.


**Q:** Can gravity assist in the exploration of cosmic phenomena like cosmic rays, which are high-energy particles originating from various sources, including supernovae and active galactic nuclei, and what can the study of cosmic rays reveal about the most energetic processes in the universe and their impact on cosmic evolution?

**A:** Gravity can contribute to the study of cosmic rays, which originate from a variety of high-energy sources. Investigating cosmic rays provides insights into the most energetic processes in the universe and their role in cosmic evolution, including their impact on the interstellar medium and the cosmic ecosystem.


**Q:** How does gravity influence the behavior of cosmic phenomena like the cosmic microwave background's polarization, which carries information about the universe's early moments, and what can the study of polarization patterns reveal about the early universe's conditions, the inflationary period, and the nature of primordial gravitational waves?

**A:** Gravity plays a role in shaping the polarization patterns of the cosmic microwave background. Studying these patterns provides information about the early universe's conditions, the inflationary period, and the existence of primordial gravitational waves. It offers a window into the universe's earliest moments.


**Q:** Can gravity assist in the investigation of the interactions between dark matter and visible matter within galaxy clusters, and how do gravitational effects reveal the presence and distribution of dark matter in these massive structures and influence their evolution?

**A:** Gravity is instrumental in studying the interactions between dark matter and visible matter within galaxy clusters. Gravitational effects help reveal the presence and distribution of dark matter in these massive structures, shedding light on their formation and evolutionary processes.

**Q:** How does gravity influence the behavior of cosmic phenomena like gravitational lensing by individual galaxies, where massive objects bend and distort the light from background objects, and what can the study of this phenomenon reveal about the distribution of mass within galaxies and the presence of dark matter?

**A:** Gravity is responsible for gravitational lensing by individual galaxies, offering insights into the distribution of mass within galaxies and the presence of dark matter. The study of these lensing effects helps astronomers map the invisible dark matter structures in galaxies.

**Q:** Can gravity assist in the understanding of cosmic microwave background (CMB) anomalies like the Axis of Evil, which represents an alignment in the CMB's temperature fluctuations, and what implications could such alignments have for our models of the universe's early moments and cosmic evolution?

**A:** Gravity is central to understanding the behavior of matter and energy in the early universe, potentially leading to anomalies like the Axis of Evil in the CMB. Investigating these alignments challenges our models of the universe's early moments and offers opportunities to refine our understanding of cosmic evolution.

**Q:** How does gravity influence the behavior of cosmic phenomena like the orbits of exoplanets within stellar systems, and what can the study of these exoplanetary systems reveal about the conditions necessary for planetary habitability and the prevalence of Earth-like worlds in the galaxy?

**A:** Gravity governs the orbits of exoplanets within stellar systems, shaping their dynamics and stability. Studying these systems provides insights into the conditions required for planetary habitability and the potential abundance of Earth-like worlds in the galaxy.

**Q:** Can gravity be harnessed to explore the nature of cosmic phenomena like cosmic strings, through their gravitational effects on light and matter, and what role could such strings play in the formation of cosmic structures and the fabric of spacetime?

**A:** Gravity can be used to explore cosmic strings through their gravitational effects on light and matter. Understanding cosmic strings may shed light on their role in the formation of cosmic structures and their influence on the fabric of spacetime itself.

# GRAVITY-ASSISTED
# TIME MANIPULATION

**Q:** How does gravity influence the behavior of cosmic phenomena like the Lagrangian points in the Earth-Sun system, where gravitational forces allow objects to maintain stable positions relative to the Earth and the Sun, and what applications do these points have in space missions, such as space telescopes and interplanetary travel?

**A:** Gravity is integral to the existence of stable Lagrangian points in the Earth-Sun system, enabling stable positions for objects relative to both celestial bodies. These points have numerous applications in space missions, including the placement of space telescopes and the optimization of interplanetary travel.

**Q:** Can gravity assist in the investigation of cosmic phenomena like the interstellar medium's turbulence, where gravitational interactions play a role in shaping the chaotic motion of cosmic gas and dust, and what can the study of turbulence reveal about the processes driving star formation and galactic evolution?

**A:** Gravity plays a role in the turbulence of the interstellar medium, influencing the chaotic motion of gas and dust. Studying this turbulence provides insights into the processes governing star formation, the evolution of galaxies, and the distribution of matter in the cosmos.

**Q:** How does gravity influence the behavior of cosmic phenomena like the orbits of binary black hole systems, where two black holes orbit each other before merging, and what can the study of these systems reveal about the existence of gravitational waves, as predicted by Einstein's theory of general relativity, and the nature of extreme gravitational environments?

**A:** Gravity is central to the orbits of binary black hole systems, leading to the emission of gravitational waves as predicted by Einstein's theory of general relativity. Studying these systems not only confirms the existence of gravitational waves but also provides insights into extreme gravitational environments and the dynamics of black hole mergers.

**Q:** Can gravity be used to probe the internal structures and composition of celestial bodies like asteroids and comets, by studying their gravitational effects on spacecraft or other objects passing nearby, and what information can such studies reveal about the origins and characteristics of these small solar system objects?

**A:** Gravity can be harnessed to probe the internal structures and composition of asteroids and comets through spacecraft encounters. These studies offer valuable information about the origins, characteristics, and potential resources of these small solar system objects.

**Q:** How does gravity influence the behavior of cosmic phenomena like the formation and motion of galactic superclusters, the largest known structures in the universe, and what can the study of superclusters reveal about the cosmic web's evolution and the interconnectedness of galaxies on a grand scale?

**A:** Gravity plays a fundamental role in the formation and motion of galactic superclusters, offering insights into the cosmic web's evolution. Studying superclusters helps us understand how galaxies are interconnected on the grandest scales and how they contribute to the structure of the universe.

**Q:** Can gravity assist in the study of cosmic phenomena like the interaction between supermassive black holes within merging galaxies, and how do these gravitational interactions shape the evolution of galaxies and the emission of powerful phenomena like gravitational waves and active galactic nuclei?

**A:** Gravity is pivotal in studying the interactions between supermassive black holes within merging galaxies. These interactions influence the evolution of galaxies and give rise to powerful phenomena such as gravitational waves and active galactic nuclei. Understanding these processes provides insights into galactic dynamics and cosmic phenomena.

**Q:** How does gravity influence the behavior of cosmic phenomena like cosmic ray propagation through interstellar and intergalactic space, and what can the study of cosmic rays reveal about the extreme environments and high-energy processes occurring in distant regions of the universe?

**A:** Gravity affects the behavior of cosmic rays as they propagate through interstellar and intergalactic space. Studying cosmic rays provides insights into the extreme environments and high-energy processes occurring in distant cosmic regions, such as supernovae and active galactic nuclei.

**Q:** Can gravity be utilized to investigate the dynamics of star clusters and globular clusters, which are dense agglomerations of stars bound by gravity, and what can the study of these clusters reveal about stellar evolution, the formation of globular clusters, and the age of the universe?

**A:** Gravity is central to understanding the dynamics of star clusters and globular clusters. Studying these clusters provides insights into stellar evolution, the formation of globular clusters, and the age of the universe, as the clusters' ages can serve as cosmic timekeepers.

Peering into the quantum weave
of spacetime, where gravity
bends knowledge.

**Q:** How does gravity influence the behavior of cosmic phenomena like the evolution of planetary atmospheres, where gravitational forces retain or lose atmospheric gases over geological timescales, and what can the study of planetary atmospheres reveal about the habitability of worlds within and beyond our solar system?

**A:** Gravity governs the evolution of planetary atmospheres, influencing whether a planet retains or loses its atmospheric gases over long periods. The study of planetary atmospheres offers crucial insights into the habitability of planets within and beyond our solar system, as atmospheres play a vital role in supporting life.

**Q:** Can gravity assist in understanding the role of tidal forces in shaping the behavior of celestial objects like moons and satellites, and how do these gravitational interactions impact the geological and orbital characteristics of these bodies within their parent systems?

**A:** Gravity is integral to the understanding of tidal forces, which shape the behavior of celestial objects such as moons and satellites. These gravitational interactions have significant impacts on the geological and orbital characteristics of these bodies within their parent systems, driving processes like tidal heating and orbital resonances.

**Q:** How does gravity influence the behavior of cosmic phenomena like the formation and evolution of protostellar disks, where planets and solar systems are born, and what can the study of these disks reveal about the processes that give rise to planetary systems and the diversity of worlds in the cosmos?

**A:** Gravity plays a critical role in the formation and evolution of protostellar disks, the birthplaces of planets and solar systems. Studying these disks provides insights into the processes that lead to planetary systems and the multitude of worlds found throughout the universe.

**Q:** Can gravity be harnessed to investigate the properties of cosmic phenomena like galactic halos, regions of dark matter that envelop galaxies, and what can the study of these halos reveal about the nature and distribution of dark matter in the universe?

**A:** Gravity can be used to study galactic halos, which are regions of dark matter surrounding galaxies. Exploring these halos provides valuable information about the nature and distribution of dark matter in the universe, contributing to our understanding of its role in cosmic structure.

**Q:** How does gravity influence the behavior of cosmic phenomena like the formation of galaxy clusters, the largest gravitationally-bound structures in the universe, and what can the study of these clusters reveal about the distribution of matter, dark matter, and the expansion rate of the universe on cosmic scales?

**A:** Gravity is the driving force behind the formation of galaxy clusters, offering insights into the distribution of matter and dark matter on cosmic scales. Studying these clusters provides valuable information about the expansion rate of the universe and the large-scale structure of the cosmos.

**Q:** Can gravity assist in the exploration of cosmic phenomena like cosmic voids, vast regions of space with low matter density, and how do gravitational interactions within these voids influence the cosmic web's overall structure and the motion of galaxies within it?

**A:** Gravity plays a role in the exploration of cosmic voids, where gravitational interactions influence the structure of the cosmic web and the motion of galaxies within it. Studying these voids offers valuable information about the large-scale distribution of matter in the universe.

**Q:** How does gravity influence the behavior of cosmic phenomena like the interactions between galaxies in galaxy groups, where several galaxies are bound together by gravity, and what can the study of these interactions reveal about the processes of galaxy mergers, star formation, and the cosmic assembly of structures?

**A:** Gravity governs the interactions between galaxies in galaxy groups, shaping processes such as galaxy mergers and star formation. Studying these interactions provides insights into the mechanisms driving the cosmic assembly of structures and the evolution of galaxies.

**Q:** Can gravity be harnessed to investigate the behavior of cosmic phenomena like gravitational instabilities in accretion disks around black holes, which lead to the formation of powerful jets of radiation and matter, and what can the study of these instabilities reveal about the extreme gravitational environments near black holes and their impact on cosmic ecosystems?

**A:** Gravity is instrumental in studying gravitational instabilities in accretion disks around black holes, leading to the formation of powerful jets. Investigating these instabilities sheds light on the extreme gravitational environments near black holes and their influence on cosmic ecosystems, including the feedback they provide to galaxies.

# Imagining a gravity-based computer...

**Q:** How does gravity influence the behavior of cosmic phenomena like the tidal interactions between galaxies in galactic clusters, where gravitational forces distort the shapes and dynamics of galaxies, and what can the study of these tidal effects reveal about the past and future of galaxies within clusters and the evolution of large-scale cosmic structures?

**A:** Gravity plays a central role in tidal interactions between galaxies in galactic clusters, distorting their shapes and dynamics. Studying these tidal effects provides insights into the past and future of galaxies within clusters and their contributions to the evolution of large-scale cosmic structures.

**Q:** Can gravity assist in understanding the role of cosmic magnetic fields in shaping the behavior of charged particles in astrophysical environments, including their interactions with cosmic radiation, the formation of magnetized plasmas, and the generation of astrophysical jets and outflows?

**A:** Gravity does contribute to understanding the behavior of charged particles in astrophysical environments, where magnetic fields play a crucial role. Investigating these interactions helps elucidate phenomena such as the formation of magnetized plasmas and the generation of astrophysical jets and outflows.

**Q:** How does gravity influence the behavior of cosmic phenomena like the interactions between planets and their moons, where gravitational forces lead to tidal effects, orbital resonances, and the sculpting of moons' surfaces, and what can the study of these interactions reveal about the histories and evolution of planetary systems in our solar system and beyond?

**A:** Gravity governs the interactions between planets and their moons, resulting in tidal effects, orbital resonances, and surface sculpting. Studying these interactions offers insights into the histories and evolution of planetary systems, both within our solar system and beyond, contributing to our understanding of exoplanetary systems.

**Q:** How does gravity influence the behavior of cosmic phenomena like the formation of massive galaxy clusters, which are the largest gravitationally-bound structures in the universe, and what can the study of these clusters reveal about the early universe's conditions, the nature of dark matter, and the cosmic web's intricate structure?

**A:** Gravity plays a pivotal role in the formation of massive galaxy clusters, shedding light on the early universe's conditions, the presence and distribution of dark matter, and the intricate structure of the cosmic web. Studying these clusters provides a window into the cosmos' evolution.

**Q:** Can gravity assist in the understanding of cosmic phenomena like cosmic microwave background (CMB) anomalies, including the hemispherical power asymmetry, where one side of the CMB has slightly more temperature fluctuations than the other, and what implications do such anomalies have for our understanding of cosmic inflation and the universe's initial conditions?

**A:** Gravity is central to understanding the behavior of matter and energy in the early universe, potentially leading to CMB anomalies like the hemispherical power asymmetry. Investigating these anomalies challenges our understanding of cosmic inflation and the universe's initial conditions, offering opportunities to refine our cosmological models.

**Q:** How does gravity influence the behavior of cosmic phenomena like the motion of galaxies within galactic superclusters, where numerous galaxy clusters are gravitationally bound together, and what can the study of these superclusters reveal about the large-scale structure of the universe and the dynamics of cosmic expansion?

**A:** Gravity governs the motion of galaxies within galactic superclusters, which are gravitationally bound conglomerates of galaxy clusters. Studying superclusters provides crucial insights into the large-scale structure of the universe and the dynamics of cosmic expansion, helping us understand how the cosmos has evolved over time.

**Q:** Can gravity be harnessed to explore the nature of cosmic phenomena like the dark matter filaments that connect galaxy clusters in the cosmic web, and what role do these filaments play in the distribution of dark matter and the formation of cosmic structures on the largest scales?

**A:** Gravity can be used to explore dark matter filaments, the cosmic threads that connect galaxy clusters in the cosmic web. Investigating these filaments offers insights into the distribution of dark matter and its role in shaping cosmic structures on the grandest scales.

**Q:** How does gravity influence the behavior of cosmic phenomena like the orbital resonances between moons and satellites within our solar system, and what can the study of these resonances reveal about the history and evolution of celestial bodies in our cosmic neighborhood?

**A:** Gravity is the driving force behind orbital resonances between moons and satellites within our solar system. Studying these resonances provides valuable insights into the history and evolution of celestial bodies in our cosmic neighborhood, offering clues about their past interactions and future trajectories.

## Precision Gravity Mapping for Subsurface or Planetary Exploration

**Q:** Can gravity assist in understanding the role of cosmic magnetic fields in shaping the behavior of charged particles within galaxies, star-forming regions, and active galactic nuclei, and what can the study of these magnetic fields reveal about the dynamics and energy processes occurring in these astrophysical environments?

**A:** Gravity does play a role in studying cosmic magnetic fields and their effects on charged particles within various astrophysical environments. Investigating these magnetic fields provides insights into the dynamics and energy processes occurring in galaxies, star-forming regions, and active galactic nuclei.

**Q:** How does gravity influence the behavior of cosmic phenomena like the interactions between stars in dense stellar clusters, where gravitational forces can lead to close encounters and even stellar mergers, and what can the study of these interactions reveal about the dynamics and evolution of star clusters and the broader galactic environment?

**A:** Gravity governs the interactions between stars in dense stellar clusters, leading to close encounters and, in some cases, stellar mergers. Studying these interactions unveils the dynamics and evolution of star clusters and their impact on the broader galactic environment.


## 7.2 Topics like "What happens if you jump on the moon?" or "How does gravity affect your health?"

**Q:** What happens if you jump on the moon?

**A:** Jumping on the Moon would be a vastly different experience from jumping on Earth due to the Moon's much lower gravity. The Moon's gravity is only about 1/6th as strong as Earth's. When you jump on the Moon, you would experience a much lower force pulling you downward, allowing you to jump much higher and stay in the air much longer than on Earth. You'd also feel lighter, as the gravitational force exerted on your body is significantly reduced. The lower gravity on the Moon is a fascinating aspect of lunar exploration, allowing astronauts to perform extraordinary leaps and explore the lunar surface in unique ways.


**Q:** How does gravity affect your health?

**A:** Gravity has a profound impact on human health, as our bodies have evolved to function in Earth's gravitational field. Here are a few key ways gravity affects our health:


*Muscle and Bone Health:* Gravity exerts a constant force on our bodies, which is essential for maintaining healthy muscles and bones. Weight-bearing activities like walking and standing stimulate bone density and muscle strength. In the absence of gravity, such as during prolonged space travel, astronauts can experience muscle atrophy and bone loss.


*Circulation:* Gravity helps blood circulate in our bodies. When we stand upright, blood is pumped against gravity from our legs to our heart. In microgravity environments, like the International Space Station (ISS), astronauts may experience fluid shifts, which can affect blood pressure and cardiovascular health.

*Balance and Coordination:* Gravity provides the necessary force for balance and coordination. In altered gravitational environments, like space, astronauts may initially experience difficulties with spatial orientation and coordination.

*Spinal Health:* Gravity places a load on our spines, helping maintain their structure. In microgravity, spinal elongation can occur, leading to back pain and discomfort.

*Overall Well-being:* Gravity's constant pull contributes to our overall sense of well-being. Extended periods in space can affect mood and sleep patterns, potentially leading to psychological stress.

**Q:** Why do objects fall when dropped?

**A:** Objects fall when dropped due to the force of gravity pulling them toward the Earth's center. This force, known as gravity, is responsible for the attraction between all objects with mass. When you release an object, gravity accelerates it downward until it reaches the ground.

**Q:** Why do we weigh less on the Moon?

**A:** We weigh less on the Moon because its gravitational force is much weaker than Earth's. The Moon's gravity is about 1/6th as strong as Earth's gravity. This means that the Moon exerts less force on our bodies, making us feel lighter and allowing us to jump higher and move more easily.

**Q:** Why do you feel weightless in an elevator when it's descending at a constant speed?

**A:** In an elevator descending at a constant speed, you feel weightless because you and the elevator are in free fall together. Gravity is still acting on you, but since both you and the elevator are falling at the same rate, there is no normal force pushing against you, making it feel like you're weightless.

**Q:** Why do astronauts float in space?

**A:** Astronauts float in space because they are in a state of continuous free fall around the Earth. This state is known as microgravity or weightlessness. While gravity still affects them and the spacecraft, they are all falling together, creating the sensation of weightlessness. It's as if they are in a perpetual state of free fall, allowing them to float and move freely in the spacecraft.

**Q:** Why does a ball follow a curved path when thrown horizontally?

**A:** A ball follows a curved path when thrown horizontally due to the combination of its horizontal motion (forward) and the force of gravity pulling it downward. As the ball moves forward, gravity acts on it, causing it to fall toward the Earth. This results in a curved trajectory known as a parabola.

**Q:** Why do we use weight instead of mass to measure our bodies?

**A:** We use weight instead of mass to measure our bodies because weight is a measure of the force of gravity acting on an object with mass. Mass, on the other hand, is a measure of the amount of matter in an object and is constant regardless of location. Since our weight varies depending on the strength of gravity (e.g., on the Moon or Earth), it provides a more practical measure for understanding how objects interact with gravity in different environments.

**Q:** Why does water flow downhill?

**A:** Water flows downhill due to the force of gravity. Gravity pulls all objects with mass, including water, toward the Earth's center. When water is at a higher elevation, it has potential energy. As it flows downhill, this potential energy is converted into kinetic energy, causing the water to move downward, following the path of least resistance.

**Q:** Why do astronauts exercise in space?

**A:** Astronauts exercise in space to counteract the negative effects of prolonged weightlessness on their muscles and bones. In microgravity, muscles and bones can weaken and atrophy because they are not subjected to the usual forces of gravity. Regular exercise, such as resistance training and cardiovascular workouts, helps maintain their physical health and reduces the risk of bone density loss and muscle atrophy.

**Q:** Why do astronauts experience muscle and bone loss in space?

**A:** Astronauts experience muscle and bone loss in space because the absence of Earth's gravity reduces the need for load-bearing and muscle work. In microgravity, the body's muscles and bones weaken due to the lack of resistance from gravity. Astronauts must exercise rigorously to counteract these effects during their missions.

**Q:** How does gravity affect the shape of celestial bodies like planets and stars?

**A:** Gravity determines the shape of celestial bodies like planets and stars. Under the influence of gravity, these objects tend to form roughly spherical shapes. Gravity pulls material toward the center, causing it to compress and shape into spheres over millions of years.

**Q:** Why do we feel gravitational force pulling us toward the center of the Earth and not in other directions?

**A:** We feel gravitational force pulling us toward the center of the Earth because the Earth's mass creates a gravitational field that exerts force in all directions. This force is strongest in the direction toward the center of the Earth, which is why we are pulled downward.

**Q:** How does gravity affect the trajectory of objects thrown into the air?

**A:** Gravity affects the trajectory of objects thrown into the air by constantly pulling them downward. When an object is thrown upward, gravity slows its ascent until it reaches its highest point (the apex) and then pulls it back down. The object follows a curved path known as a parabola due to gravity's influence.

**Q:** Why do objects of different masses fall at the same rate in a vacuum?

**A:**

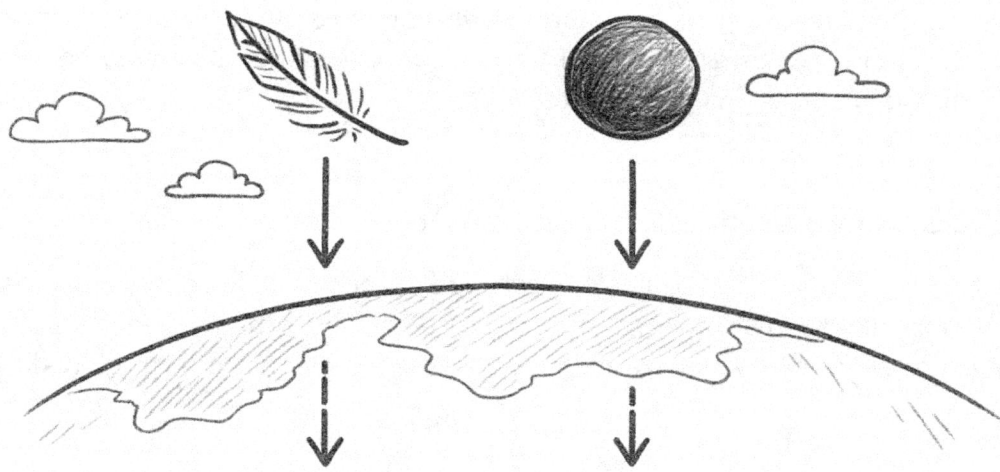

In a vacuum, objects of different masses fall at the same rate because there is no air resistance to slow them down. According to Galileo's famous experiment, all objects in a vacuum, regardless of their mass, accelerate toward the Earth's surface at the same rate, approximately 9,8 meters per second squared.

**Q:** How does gravity influence the formation of galaxies and the universe's large-scale structure?

**A:** Gravity is the driving force behind the formation of galaxies and the large-scale structure of the universe. It acts to pull matter together over cosmic distances. Gravity causes regions of higher density to attract more matter, leading to the formation of galaxies, galaxy clusters, and cosmic filaments.

**Q:** Why do astronauts appear weightless in the International Space Station (ISS)?

**A:** Astronauts appear weightless in the International Space Station (ISS) because they are in a state of continuous free fall or orbit around Earth. The ISS and everything inside it, including astronauts, are falling toward Earth while moving forward at the same rate, creating the sensation of weightlessness.

**Q:** How does gravity affect the passage of time?

**A:** Gravity affects the passage of time as described by Einstein's theory of general relativity. In areas of strong gravity, time passes more slowly compared to areas with weaker gravity. This phenomenon, known as gravitational time dilation, has been observed and measured, such as in the case of clocks on GPS satellites.

**Q:** Why do objects in space remain in motion indefinitely?

**A:** Objects in space remain in motion indefinitely due to the principle of inertia. In the vacuum of space, there is minimal resistance to motion, so objects will continue moving at a constant velocity unless acted upon by an external force. This is consistent with Newton's first law of motion.

**Q:** How does gravity affect the way we walk and move on Earth?

**A:** Gravity affects the way we walk and move on Earth by pulling us toward the ground. When we walk or move, our muscles and skeletal structure work against gravity to maintain an upright posture and control our movements. Gravity provides the necessary resistance for activities like walking and running.

**Q:** Why do astronauts experience difficulty sleeping in space?

**A:** Astronauts may experience difficulty sleeping in space due to changes in their sleep environment and the absence of Earth's natural day-night cycle. In the International Space Station (ISS), they witness 16 sunrises and sunsets each day, making it challenging to establish a regular sleep pattern. Additionally, microgravity can affect sleep quality, and astronauts often use sleep masks and earplugs to improve their sleep.

**Q:** How does gravity affect the shape of water droplets in space?

**A:**

In microgravity, gravity's influencee on the shape of water droplets is greatly diminished. On Earth, gravity causes water droplets to form spherical shapes due to surface tension. In space, without the strong pull of gravity, water droplets tend to cling together and form larger, irregular shapes.

**Q:** How does gravity affect the aging process on Earth?

**A:** Gravity plays a role in the aging process on Earth by constantly exerting force on our bodies. Over time, the effects of gravity can lead to the gradual sagging and wrinkling of skin, as well as changes in posture and joint health. These gravitational effects contribute to the visible signs of aging.

**Q:** Why do astronauts experience fluid shifts in space, and how does gravity play a role?

**A:** Astronauts in microgravity experience fluid shifts because the absence of gravity allows bodily fluids to move upward to the upper body and head. On Earth, gravity helps distribute fluids evenly throughout the body. In space, without gravity's pull, fluids accumulate in the upper body, leading to facial puffiness and pressure on the eyes.

**Q:** How does gravity affect the human cardiovascular system?

**A:** Gravity plays a crucial role in regulating blood flow in the human cardiovascular system. On Earth, gravity helps return blood from the lower extremities to the heart, preventing pooling of blood in the legs. In microgravity, this natural mechanism is disrupted, leading to changes in blood distribution and cardiovascular function in astronauts.

**Q:** How does gravity influence the behavior of gases in Earth's atmosphere?

**A:** Gravity influences the behavior of gases in Earth's atmosphere by creating a pressure gradient. Near the Earth's surface, the weight of the atmosphere above exerts pressure, with higher pressure at lower altitudes. This pressure gradient is responsible for many atmospheric phenomena, including wind patterns and weather systems.

**Q:** Why does water flow downhill and not uphill?

**A:** Water flows downhill due to the force of gravity. Gravity pulls water toward the center of the Earth, causing it to seek the lowest point available. This gravitational pull is responsible for rivers flowing from highlands to lowlands, ensuring water always moves downhill.

**Q:** Why do objects feel heavier at the bottom of a swimming pool or in deep water?

**A:** Objects feel heavier in deep water or at the bottom of a swimming pool because the pressure increases with depth. This increased pressure is due to the weight of the water above, and it adds to the sensation of weight when you hold objects underwater. Gravity still acts on these objects, but the added pressure from the water makes them feel heavier.

**Q:** How does gravity affect the shape and stability of structures like buildings and bridges?

**A:** Gravity plays a crucial role in the design and stability of structures like buildings and bridges. Engineers must consider the gravitational force acting on these structures to ensure they can support their own weight and the loads they will bear over time. Gravity helps anchor these structures to the ground and keeps them standing.

**Q:** Why do we use ramps or inclined planes to move heavy objects?

**A:**

Ramps or inclined planes are used to move heavy objects more easily because they reduce the effects of gravity. By spreading the effort over a longer distance, it requires less force to lift an object up a ramp compared to lifting it vertically. This makes it more practical to move heavy loads.

**Q:** How does gravity affect the balance and stability of vehicles like cars and bicycles?

**A:** Gravity affects the balance and stability of vehicles like cars and bicycles by providing the necessary force to keep them grounded. For bicycles, gravity keeps the wheels on the road, helping maintain stability. In cars, gravity prevents them from lifting off the ground as they move. Proper weight distribution is critical for vehicle stability.

**Q:** Why do objects fall when we release them, but they stay in place on a table?

**A:** Objects fall when we release them because they are subject to the force of gravity, which pulls them toward the center of the Earth. When placed on a table or a surface, the table exerts an equal and opposite force (a normal force) that counteracts gravity, keeping the object in place.

**Q:** How does gravity influence our balance and coordination when we walk or run?

**A:** Gravity influences our balance and coordination when we walk or run by pulling us downward, providing the necessary force for maintaining an upright posture. Our muscles and nervous system work together to adjust our movements and maintain balance as we navigate different terrains.

**Q:** Why do objects sink or float in water?

**A:** Whether an object sinks or floats in water depends on its density relative to the density of water. Objects with higher density than water (e.g., rocks) will sink because gravity pulls them downward. Objects with lower density (e.g., wood or a plastic bottle) will float because they are less dense than water and buoyant forces counteract gravity.

**Q:** Why do we feel lighter when we jump on a trampoline?

**A:** When you jump on a trampoline, the trampoline's springs provide an upward force that counters gravity, making you feel lighter as you bounce. This temporary sensation of reduced weight is due to the opposing forces at play.

**Q:** How does gravity influence the way we walk up and down stairs?

**A:** Gravity affects the way we walk up and down stairs by requiring more effort when ascending and helping us descend safely. When climbing stairs, we must overcome gravity's pull to reach a higher elevation, while descending stairs allows gravity to assist us in controlling our descent.

**Q:** Why do we need to use weights or resistance in strength training exercises?

**A:**

We use weights or resistance in strength training exercises to work against the force of gravity. These exercises require muscles to exert force to lift or move the weights against gravity's pull, which helps build muscle strength.

**Q:** How does gravity impact the way we breathe and the circulation of blood in our bodies?

**A:** Gravity influences the way we breathe and circulate blood by helping move blood and air downward in our bodies. Gravity assists in returning blood from our legs to our heart and ensures that oxygen-rich air flows into our lungs when we inhale.

**Q:** Why do we need to maintain proper posture when sitting and standing?

**A:** Proper posture is essential when sitting and standing because it helps counteract the effects of gravity on our bodies. Maintaining good posture distributes the force of gravity evenly on our muscles and skeletal structure, reducing strain and promoting long-term health.

**Q:** How does gravity affect our digestive system and the passage of food through our bodies?

**A:** Gravity plays a role in moving food through our digestive system by aiding in the peristaltic movement of the digestive tract. It helps food and waste travel in the desired direction, from the mouth to the stomach and eventually out of the body.

**Q:** Why does pouring liquid from a container require tilting it downward?

**A:** Pouring liquid from a container requires tilting it downward because gravity helps the liquid flow out. Tilting the container changes the angle and allows gravity to pull the liquid toward the opening, enabling a smoother pour.

**Q:** How does gravity influence our sense of balance when we stand on one leg or perform yoga poses?

**A:** Gravity influences our sense of balance when we stand on one leg or perform yoga poses by constantly pulling us toward the Earth. These balancing activities require muscles and the body's proprioceptive system to adjust and maintain stability against gravity's force.

**Q:** Why do liquids fill the bottom of a container evenly when it's placed on a flat surface?

**A:** Liquids fill the bottom of a container evenly due to the influence of gravity. Gravity pulls the liquid downward, causing it to settle uniformly at the lowest point in the container, which is the bottom.

**Q:** How does gravity affect the way we sleep, and why do we prefer lying down horizontally when resting?

**A:**

Gravity affects the way we sleep by making it more comfortable to lie down horizontally. When we sleep in a horizontal position, gravity evenly distributes our body weight across a larger surface area, reducing pressure points and promoting better rest.

**Q:** Why do heavy objects tend to stay in place when we set them down on a table, while lighter objects may move more easily?

**A:** Heavy objects tend to stay in place on a table because their greater mass provides resistance to being moved by external forces, including gravity. Lighter objects may be more easily affected by small forces and could move if disturbed.

**Q:** How does gravity influence the growth of plants, and why do plants grow upwards toward the sky?

**A:** Gravity influences plant growth by pulling the roots downward and helping the plant maintain stability. Plants grow upwards toward the sky through a process called phototropism, where they reach for sources of light while gravity ensures their roots remain anchored in the soil.

**Q:** Why do we use weights on balloons to keep them grounded, and how does gravity affect their flight?

**A:** We use weights on balloons to counteract gravity's upward pull. Without these weights, helium-filled balloons would rise because helium is less dense than the surrounding air. The addition of weights ensures they remain grounded and don't float away.


**Q:** How does gravity affect the way we cook and prepare food in the kitchen?

**A:** Gravity affects the way we cook and prepare food in the kitchen by causing liquids to flow downward when poured, guiding the direction of falling objects, and helping to evenly distribute heat in pans and pots.


**Q:** Why do we use ramps for loading and unloading heavy objects onto trucks and trailers?

**A:** Ramps are used for loading and unloading heavy objects onto trucks and trailers because they reduce the effects of gravity and make it easier to move objects between different elevations. The inclined surface allows for a gradual ascent or descent, minimizing the force needed to move heavy loads.


**Q:** How does gravity affect our perception of time when we're waiting or experiencing time passing slowly?

**A:** Gravity does not directly affect our perception of time. However, when we're waiting or experiencing time passing slowly, it's usually due to our subjective perception of events and our mental state, not gravity.


**Q:** Why does water flow downhill when we're washing dishes or taking a shower?

**A:** Water flows downhill during activities like washing dishes or taking a shower because of gravity. Gravity pulls the water down toward the lowest point, guiding its path along pipes and drains to ensure it flows away.


**Q:** How does gravity influence the way we pack our bags for a trip?

**A:** Gravity influences the way we pack our bags by making us mindful of weight distribution. Placing heavier items at the bottom of the bag helps lower its center of gravity and makes it more stable and comfortable to carry.

**Q:** Why do we feel a sense of weight when we hold something heavy in our hands?

**A:** We feel a sense of weight when holding something heavy because gravity is pulling the object downward, and our muscles and nerves sense and respond to this force. This sensation helps us gauge the object's mass and adjust our grip and posture accordingly.

**Q:** How does gravity impact the way we design and construct buildings and structures on Earth?

**A:** Architects and engineers consider gravity's constant downward force when determining load-bearing capacities, ensuring structures can safely support their own weight and external loads.

## Gravity significantly influences architectural and engineering decisions when designing buildings and structures.

**Q:** Why do roller coasters have steep drops and loops, and how does gravity play a role in creating thrilling rides?

**A:** Roller coasters have steep drops and loops to provide thrilling experiences for riders. Gravity is a key component in roller coaster design, as it accelerates the coaster downward during drops and generates the forces that create exhilarating sensations, such as airtime and G-forces.

**Q:** How does gravity impact the way we navigate hilly or mountainous terrain while hiking or driving?

**A:** Gravity influences the way we navigate hilly or mountainous terrain by causing changes in elevation. When hiking or driving uphill, we must work against gravity's pull, requiring more effort. Conversely, when going downhill, gravity aids in the descent but necessitates careful control to avoid accelerating too quickly.

**Q:** Why does gravity cause objects to fall straight down rather than at an angle when released from a height?

**A:** Gravity causes objects to fall straight down rather than at an angle when released from a height because gravity acts vertically toward the center of the Earth. Objects follow a vertical path due to the direct influence of gravity, without any horizontal forces acting on them.

**Q:** How does gravity affect the way we interact with our surroundings, from picking up objects to climbing stairs?

**A:** Gravity is a fundamental aspect of our interactions with the environment. It provides the necessary resistance and forces that enable us to perform everyday activities, such as lifting objects, walking, running, and climbing stairs. Gravity plays a central role in maintaining our connection to the Earth's surface.

**Q:** Why do liquids flow to the bottom of a glass when poured, and how does gravity affect this process?

**A:** Liquids flow to the bottom of a glass when poured because of gravity. Gravity exerts a force that pulls the liquid downward, filling the container from the bottom up. It's this gravitational force that ensures the liquid takes the shape of the container.

**Q:** How does gravity affect the way we perform activities like jumping, running, and dancing?

**A:** Gravity is a constant force that influences our movements during activities like jumping, running, and dancing. It affects how high we can jump, how fast we can run, and how gracefully we can dance by determining the resistance we encounter and the energy required for these actions.

**Q:** Why does gravity play a role in how we perceive weight and heaviness when holding objects?

**A:** Gravity plays a significant role in how we perceive weight and heaviness when holding objects. The gravitational force pulling objects toward the Earth affects our sense of touch and the pressure we feel when holding something. Heavier objects exert more gravitational force on us, leading to a stronger sensation of weight.

**Q:** How does gravity influence our posture and spinal health as we stand and sit throughout the day?

**A:** Gravity affects our posture and spinal health as we stand and sit by constantly pulling us downward. Maintaining good posture is essential to counteract gravity's force and prevent poor spinal alignment, which can lead to discomfort and back problems over time.

**Q:** Why do we use weights or resistance in exercise to build muscle strength, and how does gravity factor into this?

**A:** We use weights or resistance in exercise to build muscle strength because they provide an opposing force against gravity. Lifting weights or engaging in resistance exercises requires muscles to overcome the gravitational force acting on the weights, leading to muscle growth and increased strength.

**Q:** How does gravity influence the way we experience g-forces when driving in a car or riding a roller coaster?

**A:** Gravity influences the way we experience g-forces when driving in a car or riding a roller coaster. When a vehicle accelerates, decelerates, or changes direction, these forces act in conjunction with gravity, affecting how we feel during these maneuvers. For example, when a car accelerates, we feel pushed back into the seat due to the combination of acceleration and gravity.

**Q:** How does gravity contribute to the natural flow of fluids in our bodies, such as blood circulation and the movement of digestive fluids?

**A:** Gravity plays a key role in the natural flow of fluids in our bodies. It assists in blood circulation by aiding the return of blood from our limbs to the heart. Gravity also facilitates the movement of digestive fluids, helping to transport food and waste through our digestive system.

**Q:** Why do we need to consider gravity when arranging furniture and objects in our homes to ensure stability and safety?

**A:** We need to consider gravity when arranging furniture and objects in our homes to ensure stability and safety. Gravity can topple unbalanced objects, so arranging them securely helps prevent accidents. Proper weight distribution and secure anchoring prevent objects from falling or shifting unexpectedly.

**Q:** Why does gravity cause our muscles to work harder when we climb stairs or hike uphill?

**A:** Gravity makes our muscles work harder when climbing stairs or hiking uphill because it opposes our upward movement. To overcome gravity, our leg muscles must exert more force to lift our body weight and ascend to a higher elevation.

**Q:** How does gravity impact the way we control vehicles like bicycles and cars when going downhill?

**A:** Gravity influences the way we control vehicles like bicycles and cars when going downhill by increasing their speed. Gravity pulls these vehicles downward, and without proper braking or control, they can accelerate rapidly. This requires careful handling to maintain safety.

**Q:** Why do we experience a sensation of "weightlessness" when riding a free-fall amusement park ride, and how does gravity play a role in these experiences?

**A:** We experience a sensation of "weightlessness" on free-fall amusement park rides because the ride momentarily counteracts the force of gravity. As the ride drops, it and its passengers are in free fall, leading to a sensation of weightlessness until gravity begins to slow the descent.

**Q:** Why do astronauts experience "weightlessness" in space, and how does it affect their bodies and daily activities on the International Space Station (ISS)?

Astronauts experience 'weightlessness' in space because they are in a state of free fall around the Earth. This condition is often called microgravity.

**A:** In the ISS, astronauts float and objects appear to be weightless because they are all falling at the same rate due to gravity, creating a sensation of weightlessness. This affects daily activities, such as eating, sleeping, and conducting experiments, which must be adapted to a microgravity environment.

**Q:** How does gravity influence the shape and structure of fruits and vegetables on plants, and why do they tend to grow downward or toward the Earth?

**A:** Gravity influences the shape and structure of fruits and vegetables on plants by causing them to grow downward or toward the Earth. This is because plants use gravity as a cue for their orientation—roots grow down into the soil to access water and nutrients, while stems and fruits grow upward, against gravity, to reach for sunlight.

**Q:** How does gravity influence the way we perceive distances when looking at objects on the horizon or at a distance?

**A:** Gravity doesn't directly influence the way we perceive distances when looking at objects on the horizon or at a distance. However, it plays a role in how light travels through the atmosphere. Atmospheric refraction, influenced by gravity, can cause distant objects to appear slightly elevated or distorted.

**Q:** Why does gravity make it challenging to perform certain physical activities, such as balancing on one leg or doing handstands?

**A:** Gravity makes it challenging to perform certain physical activities, like balancing on one leg or doing handstands, because it constantly pulls us toward the Earth. These activities require countering the force of gravity to maintain balance or inversion, making them physically demanding.

**Q:** How does gravity affect our bodily fluids and pressure distribution when we're in different positions, such as standing, sitting, or lying down?

**A:** Gravity affects the distribution of bodily fluids and pressure when we change positions. For example, when standing, gravity pulls blood and fluids downward, causing an increase in pressure in the lower extremities. When lying down, pressure is distributed more evenly.

**Q:** Why do we use ramps for loading and unloading heavy objects onto trucks and trailers, and how does gravity factor into this process?

**A:** We use ramps for loading and unloading heavy objects onto trucks and trailers to reduce the effects of gravity. The inclined surface of the ramp allows for a gradual ascent or descent, minimizing the force needed to move heavy loads against gravity.

**Q:** How does gravity influence the way we pour and mix ingredients when cooking, and why is it important in the culinary world?

**A:** Gravity influences the way we pour and mix ingredients when cooking by guiding the flow of liquids and ensuring even distribution. It's crucial in culinary processes to maintain consistency and prevent uneven mixing or spills.

**Q:** Why does gravity play a role in the formation of ocean waves and their impact on coastal areas?

**A:** Gravity plays a key role in the formation of ocean waves. Wind-driven friction and pressure differences on the ocean's surface create waves that are influenced by gravity. These waves travel across the ocean and can have significant impacts on coastal areas, including erosion and the creation of surfable waves.

Gravity plays a key role in the formation of ocean waves.

**Q:** How does gravity affect the way we sit in chairs, and why do chairs have a specific design to accommodate our bodies and gravity?

**A:** Gravity affects the way we sit in chairs by pulling us downward. Chairs are designed to accommodate our bodies and gravity's influence by providing support and comfort. Ergonomic chair designs aim to distribute our weight evenly and promote good posture to counteract gravity's effects.

**Q:** Why do we feel pressure in our feet when standing for extended periods, and how does gravity contribute to this sensation?

**A:** We feel pressure in our feet when standing for extended periods because of gravity. Gravity continuously pulls us downward, and when standing, our body weight exerts pressure on our feet. This sensation is relieved when we sit or lie down.

**Q:** How does gravity influence the way we interact with liquid-filled containers, such as drinking from a cup or pouring from a pitcher?

**A:** Gravity influences the way we interact with liquid-filled containers by causing liquids to flow in response to gravitational force. When drinking from a cup or pouring from a pitcher, gravity guides the liquid's movement, ensuring it flows in the desired direction.

**Q:** Why do we use gravity-based systems like siphons and drains in various applications, such as transferring liquids or emptying containers?

**A:** We use gravity-based systems like siphons and drains in various applications to harness the force of gravity for efficient liquid transfer or emptying of containers. These systems rely on gravity to create flow and move liquids without the need for external power sources.

**Q:** How does gravity affect the way we use tools like plumb bobs and levels for construction and alignment purposes?

**A:** Gravity plays a central role in the operation of tools like plumb bobs and levels. Plumb bobs hang vertically due to gravity, helping to establish precise vertical reference lines. Levels rely on the horizontal reference provided by gravity to ensure surfaces are flat and aligned.

**Q:** How does gravity impact extreme sports like skateboarding and BMX biking, and why is it essential for performing tricks?

# Gravity causes projectiles to follow a parabolic trajectory.

**A:** Gravity is a key factor in extreme sports like skateboarding and BMX biking. It provides the force necessary for performing aerial tricks and stunts. Understanding gravity's effects helps athletes calculate the timing and force required for successful maneuvers.

**Q:** How does gravity affect the trajectory and physics of projectiles like basketballs or soccer balls when they are kicked or thrown?

**A:** Gravity plays a significant role in determining the trajectory of projectiles like basketballs or soccer balls. It causes them to follow a curved path, known as a parabolic trajectory. Understanding gravity helps athletes predict and control the path of the ball during play.

**Q:** Why do astronauts experience physical changes in space due to the absence of gravity, and how does this affect their bodies and health during long missions?

**A:** Astronauts experience physical changes in space because of the absence of gravity. This leads to muscle atrophy, bone density loss, and other health effects. Understanding these changes is crucial for planning long-duration space missions, such as those to Mars.

**Q:** What is the concept of "microgravity," and how is it simulated on Earth for scientific research and astronaut training?

**A:** Microgravity refers to a condition where the force of gravity is greatly reduced, as experienced in space. Scientists simulate microgravity on Earth using drop towers, parabolic flight maneuvers, and underwater training to conduct experiments and train astronauts for space missions.

**Q:** How does gravity affect the behavior of fluids like water in space, and why is this important for future space exploration and habitation?

**A:** Gravity has a significant impact on fluid behavior, and understanding this is vital for space exploration. In microgravity, fluids behave differently, making it essential to develop systems for water and waste management, as well as propulsion, for future space habitats.

**Q:** Why do objects in orbit around Earth, such as satellites, experience microgravity, and how is this utilized in satellite technology and communication?

**A:** Objects in orbit around Earth, including satellites, experience microgravity because they are in a constant state of free fall. This state is utilized in satellite technology, allowing satellites to remain in a stable orbit and provide services such as communication, navigation, and weather forecasting.

**Q:** How does gravity influence the formation of celestial bodies like stars and planets, and what role does it play in the life cycle of these objects?

**A:** Gravity is the driving force behind the formation of celestial bodies like stars and planets. It causes gas and dust in space to clump together and eventually form these objects. Gravity also governs their life cycles, including processes like nuclear fusion in stars.

**Q:** Why is the concept of "zero gravity" misleading, and what is the actual state of gravity experienced by astronauts in low Earth orbit?

**A:** The term "zero gravity" is misleading because gravity is still present in low Earth orbit. Astronauts aboard the International Space Station, for example, experience about 90% of Earth's gravity. They appear weightless due to the continuous free fall they are in around the Earth.

**Q:** What would happen if you jumped off a diving board on the Moon?

**A:**

If you jumped off a diving board on the Moon, you would soar into the air and stay afloat for a much longer time compared to Earth because the Moon has weaker gravity.

**Q:** Can you name the planet with the strongest gravity in our solar system?

**A:** Jupiter has the strongest gravity of any planet in our solar system. If you weighed 100 pounds on Earth, you would weigh about 240 pounds on Jupiter!

**Q:** Why do astronauts on the International Space Station (ISS) float around inside instead of falling to the floor?

**A:** Astronauts float inside the ISS because they are in constant free fall around the Earth. This creates the sensation of weightlessness, allowing them to float around as if there's no gravity.

**Q:** How does gravity help shape the orbits of planets in our solar system?

**A:** Gravity is like an invisible string that holds planets in orbit around the Sun. The balance between the planet's forward motion and the Sun's gravity keeps them moving in their paths.

**Q:** Did you know that the famous physicist Isaac Newton discovered the laws of gravity when an apple fell on his head?

**A:** Although it's a popular story, it's likely not true that an apple fell on Newton's head. However, Newton did observe an apple falling to the ground, which inspired his thoughts on gravity.

**Q:** Can you explain why astronauts' muscles and bones weaken in space, even though they exercise regularly?

**A:** Astronauts' muscles and bones weaken in space because they don't have to work against gravity. Our bodies need gravity to keep muscles and bones strong, so in microgravity, they become weaker.

**Q:** What would happen if Earth's gravity suddenly became half as strong? How might it affect us?

**A:** If Earth's gravity became half as strong, we would weigh half as much, and everything would feel lighter. We might be able to jump higher, but it could also cause some problems, like objects floating away more easily.

**Q:** Did you know that the force of gravity between two objects depends on their mass and distance? The bigger the mass and the closer the objects are, the stronger the gravitational force.

**A:** This is why planets like Earth have strong gravity – they have a large mass, and we're very close to them.

**Q:** How do astronauts eat and drink in space without making a mess?

**A:** Astronauts use special containers and utensils designed to prevent food and liquids from floating away. They have to be careful and control their movements to avoid making a mess!

**Q:** Why do objects fall at the same rate regardless of their mass when dropped from the same height on Earth?

**A:** Objects fall at the same rate on Earth because gravity affects them equally, causing them to accelerate at the same rate (about 9.8 meters per second squared).

**Q:** Can you name the scientist who first proposed the theory of general relativity, which revolutionized our understanding of gravity?

**A:** The scientist who proposed the theory of general relativity was Albert Einstein. His theory described gravity as the warping of spacetime by massive objects.

**Q:** What happens to your weight on a mountaintop compared to when you're at sea level, and why does it change?

**A:** Your weight is slightly less on a mountaintop than at sea level because the force of gravity is slightly weaker farther from Earth's center. As you move higher up, you get closer to the center of the Earth, where gravity is stronger.

**Q:** Why do astronauts wear spacesuits when they go on spacewalks, and how do these suits help them with gravity-related challenges in space?

**A:** Astronauts wear spacesuits to protect themselves from the vacuum of space and extreme temperatures. The suits also provide oxygen and help maintain their body pressure, crucial for survival in the absence of atmospheric pressure and gravity.

**Q:** Can you explain the concept of "escape velocity" and why it's essential for spacecraft leaving Earth's atmosphere?

**A:** Escape velocity is the minimum speed required for an object to break free from a celestial body's gravitational pull. It's crucial for spacecraft to achieve this speed to leave Earth's atmosphere and reach space.

**Q:** How does gravity influence the formation of galaxies and the way they hold stars and planets together within them?

**A:** Gravity is responsible for the formation of galaxies by pulling together vast clouds of gas and dust. It also keeps stars and planets in their orbits within galaxies, maintaining the structure of these cosmic systems.

**Q:** Why does the Moon have lower gravity than Earth, and how does this affect the way things behave on its surface?

**A:** The Moon has lower gravity than Earth because it is smaller and less massive. This lower gravity makes objects on the Moon weigh much less than they do on Earth, allowing astronauts to jump higher and cover more ground with each step.

**Q:** What would happen if Earth's gravity suddenly doubled? How might it impact our daily lives and the planet itself?

**A:** If Earth's gravity were to double, everything would weigh twice as much. It could lead to structural problems with buildings and infrastructure and make daily activities more challenging.

**Q:** Did you know that you're constantly pulling on objects with your own gravity, just like Earth does with us? However, your gravity is incredibly weak because you're much smaller than Earth!

**A:** This is why you don't notice any gravitational effects on everyday objects.

**Q:** How does gravity play a role in the formation of raindrops, and why do raindrops have different sizes?

**A:** Gravity affects raindrop formation by pulling water droplets in clouds together. Raindrops come in various sizes because they grow as they collide with other droplets and become too heavy for the air to support.

**Q:** Can you explain the concept of "microgravity" in space and how it allows astronauts to conduct experiments that wouldn't be possible on Earth?

**A:** Microgravity is a condition of very weak gravity, like what astronauts experience in space. It enables experiments where the effects of gravity are minimized, allowing scientists to study phenomena not possible in Earth's gravity.

**Q:** How does gravity influence the tides on Earth, and why do we have high and low tides each day?

**A:** Gravity from the Moon and the Sun causes the tides on Earth. High tides occur when these gravitational forces align, creating a "bulge" in the ocean. Low tides happen when they are at right angles, causing a drop in water levels.

**Q:** Why do astronauts have to exercise regularly in space, and how does this help counteract the effects of prolonged microgravity on their muscles and bones?

**A:** Astronauts exercise in space to prevent muscle and bone loss due to microgravity. Weightlifting and cardiovascular activities help maintain their health during long missions.

**Q:** What would happen if you dug a tunnel straight through the Earth from one side to the other, and how would gravity affect you inside the tunnel?

A: If you could dig a tunnel through the Earth, gravity would gradually decrease as you descended into the tunnel. At the center of the Earth, you would experience weightlessness, and then gravity would gradually increase as you approached the other side.

# Gravity decreases to zero deep underground.

**Q:** How does gravity influence the orbits of artificial satellites, such as those used for communication and navigation, and why do they need to be carefully positioned in space?

**A:** Gravity determines the orbits of artificial satellites, and they must be positioned precisely to stay in their desired orbits. This precision ensures that satellites can provide continuous services like GPS and satellite TV.


**Q:** Can you explain why astronauts on the Moon had to hop instead of walk, and how the Moon's lower gravity influenced their movements?

**A:** Astronauts on the Moon had to hop because of its lower gravity. Their suits and equipment were heavy, making walking difficult. Hopping allowed them to cover more ground with less effort.


**Q:** How does gravity affect the shape and motion of galaxies in the universe, and why do galaxies often form spiral or elliptical shapes?

**A:** Gravity influences the shape and motion of galaxies by pulling stars and gas together. Spiral galaxies have a rotating disk shape, while elliptical galaxies are more spherical due to different gravitational interactions.


**Q:** What would happen if you were inside a spaceship that suddenly stopped accelerating in deep space? Would you float or feel a force like gravity?

**A:** If a spaceship stopped accelerating in deep space, you would feel like you were floating inside because there would be no force acting like gravity. This is similar to the feeling of weightlessness in space.


**Q:** Did you know that Earth's gravity is slightly weaker at the equator compared to the poles?

**A:** This is because the Earth is not a perfect sphere; it's slightly flattened at the poles and bulging at the equator! This variation in gravity is why some locations have different weights for the same mass.

**Q:** How does gravity affect the shape of galaxies, and what are some of the different types of galaxies found in the universe?

**A:** Gravity shapes galaxies by pulling stars and gas into various configurations. Some common galaxy types include spiral galaxies (with arms), elliptical galaxies (oval-shaped), and irregular galaxies (with no specific shape).

**Q:** Can you explain why objects in orbit, like the International Space Station (ISS), appear to be in constant free fall, and how this relates to the concept of microgravity?

**A:** Objects in orbit, like the ISS, are in a state of constant free fall around the Earth. This creates a sensation of weightlessness or microgravity because everything inside the orbiting object is falling at the same rate, giving the appearance of floating.

Objects in orbit experience a
sensation of weightlessness.

**Q:** How does gravity influence the behavior of comets and their orbits in the solar system, and why do comets have long tails when they approach the Sun?

**A:** Gravity from the Sun affects comets by pulling them toward it. When comets approach the Sun, they heat up, releasing gas and dust that form a bright tail. Gravity plays a role in shaping and curving these tails.

**Q:** What would happen if you could stand on the surface of a neutron star, which has incredibly strong gravity due to its high density?

**A:** If you could somehow stand on the surface of a neutron star, you would experience an incredibly strong gravitational force that would crush you instantly. Neutron stars are so dense that a single teaspoon of their material would weigh as much as a mountain on Earth.

**Q:** How does gravity affect the atmosphere of a planet, and why do some planets have thicker atmospheres than others?

**A:** Gravity holds a planet's atmosphere in place. The thickness of an atmosphere depends on factors like a planet's size, composition, and temperature. Larger planets with stronger gravity can retain thicker atmospheres.

**Q:** Why do astronauts have to wear specially designed spacesuits in space, and how do these suits help them survive in the vacuum of space?

**A:** Astronauts wear spacesuits in space to protect themselves from the vacuum and extreme temperatures. These suits provide oxygen, maintain body pressure, and shield astronauts from harmful radiation.

**Q:** How does gravity influence the shape and behavior of galaxies in the universe, and why do galaxies cluster together in vast structures?

**A:** Gravity is responsible for the clustering of galaxies into structures like galaxy clusters and superclusters. It pulls galaxies toward one another, leading to the formation of these vast cosmic structures.

**Q:** What would happen if you jumped off a skyscraper on Mars, which has weaker gravity than Earth? Would you fall more slowly or quickly, and how would it affect your landing?

**A:** If you jumped off a skyscraper on Mars, you would fall more slowly due to its weaker gravity. When you landed, you would experience a gentler impact compared to Earth, although safety precautions would still be necessary.

**Q:** What would happen if you were in a spacecraft that got too close to a black hole, and how would the immense gravitational pull affect you and the spacecraft?

**A:** If a spacecraft got too close to a black hole, the intense gravitational pull would cause extreme tidal forces, stretching and tearing the spacecraft apart. It's an incredibly dangerous scenario due to the immense gravitational forces involved.

**Q:** How does gravity affect the way celestial bodies like moons and asteroids move through space, and why do they follow specific orbits around larger objects like planets?

**A:** Gravity determines the orbits of moons and asteroids around larger celestial bodies. These objects move in orbits because they are continually falling toward the more massive body while also moving forward tangentially, creating a balanced path.

**Q:** Why do astronauts float inside the International Space Station (ISS), and what happens to everyday objects like water and food in microgravity?

**A:** Astronauts float inside the ISS because they are in a state of continuous free fall around Earth, creating the sensation of weightlessness. In microgravity, everyday objects like water and food form floating globules rather than behaving as they do on Earth.

**Q:** How does gravity influence the behavior of gases in our atmosphere, leading to phenomena like air pressure and weather patterns?

**A:** Gravity keeps Earth's atmosphere in place, creating air pressure. Gravity also plays a role in shaping weather patterns by causing air to move from areas of high pressure to low pressure, which drives winds and weather systems.

**Q:** What is a gravitational assist, and how do spacecraft use the gravity of planets and other celestial bodies to gain speed and change their trajectories in space exploration missions?

**A:** A gravitational assist is a technique used by spacecraft to gain speed and change direction by passing close to a celestial body. By using a planet's gravity, spacecraft can save fuel and reach their destinations more efficiently.

**Q:** Can you explain the concept of time dilation in the context of general relativity, and how does gravity affect the passage of time near massive objects like black holes?

**A:** In general relativity, time dilation occurs when gravity is stronger or weaker in different locations. Near massive objects like black holes, gravity is stronger, causing time to pass more slowly. This phenomenon has been observed and confirmed in experiments.

**Q:** How do animals on Earth, such as birds and insects, use gravity to navigate during migration, and what role does gravity play in their ability to sense direction and altitude?

**A:** Animals like birds and insects use gravity as a reference for navigation during migration. They can sense changes in altitude and direction relative to the Earth's gravitational pull, helping them find their way over long distances.

## Animals like birds and insects use gravity as a reference for navigation for during migration.

**Q:** What would happen if Earth's gravity suddenly disappeared, even for a brief moment? How might it affect everything on the planet?

**A:** If Earth's gravity were to suddenly disappear, everything not anchored to the ground would float into space. Our atmosphere would disperse, and life as we know it would be unsustainable. Gravity is crucial for maintaining Earth's stability.

**Q:** Did you know that the strength of gravity varies not only on different planets but also on different parts of the same planet? For example, a mountain has slightly weaker gravity than the nearby plains due to its greater distance from Earth's center!

**A:** This variation in gravity is known as "gravity anomalies."

**Q:** How does gravity play a role in the phenomenon of ocean tides, and why do we experience two high tides and two low tides each day?

**A:** Gravity from the Moon and the Sun creates tidal forces that cause ocean tides. We experience two high tides and two low tides because Earth rotates through these tidal bulges twice a day.

**Q:** Can you explain the concept of "weightlessness" experienced by astronauts in space and how it differs from being in a microgravity environment?

**A:** Weightlessness, often called "zero gravity," is the sensation of not feeling the force of gravity. Astronauts experience weightlessness in microgravity environments because they are in a constant state of free fall, but they still have mass and are affected by gravity.

**Q:** How does gravity influence the orbits of natural satellites (moons) around planets, and why do some planets have multiple moons while others have none?

**A:** Gravity determines the orbits of moons around planets. Planets with multiple moons have captured or formed them over time due to gravitational interactions, while others may not have captured any.

**Q:** Why do astronauts need to wear special shoes with Velcro in microgravity environments like the ISS, and how does this help them stay grounded and perform tasks?

**A:** Astronauts wear shoes with Velcro to anchor themselves to surfaces and prevent floating in microgravity. Velcro sticks to surfaces, allowing them to push off and move around the spacecraft.

**Q:** How does gravity influence the motion of objects on a roller coaster, and what role does it play in the thrilling drops and loops riders experience?

**A:** Gravity provides the force that pulls roller coaster cars downhill, creating acceleration. It's responsible for the exhilarating drops and loops as the coaster's design takes advantage of gravity's effects.

**Q:** Can you explain the concept of "event horizon" in the context of black holes and how it relates to the point of no return for objects falling into a black hole?

## The event horizon is the boundary surrounding a black hole beyond which nothing, not even light, can escape.

**A:** The event horizon is the boundary surrounding a black hole beyond which nothing, not even light, can escape. It marks the point of no return for objects falling into a black hole, as they are trapped by the immense gravitational pull.

**Q:** How does gravity impact the way planets and other celestial bodies in our solar system form and evolve over time, shaping their surfaces and geological features?

**A:** Gravity plays a key role in shaping planetary surfaces by influencing processes like erosion, tectonics, and the formation of features such as mountains, valleys, and craters.

**Q:** Did you know that astronauts become taller in space due to the absence of gravity compressing their spines? They return to their normal height upon returning to Earth!

**A:** This phenomenon is temporary and related to the spine's flexibility.

**Q:** What would happen if you tried to fly like a bird by flapping your arms? How does this relate to the limitations of human flight compared to birds and other animals?

**A:** Human flight by flapping arms is not feasible due to the limitations of our muscle power compared to the wingspan and muscle efficiency of birds. Gravity significantly affects our ability to fly.

**Q:** How does gravity influence the way celestial objects like stars are born, evolve, and eventually end their lives, and what role does nuclear fusion play in this process?

**A:** Gravity initiates the collapse of gas and dust clouds, leading to star formation. Gravity also causes the compression and heating necessary for nuclear fusion, which powers stars throughout their lifecycles.

# 8

# Fun Experiments You Can Try at Home

"**The good thing about science is that it's true whether or not you believe in it.**"
— *Neil deGrasse Tyson*

Explore the fascinating world of gravity through hands-on experiments that you can easily try at home. From investigating the behavior of falling objects to creating pendulum swings and launching straw rockets, these experiments offer a captivating way to observe and understand the force that shapes our everyday lives. Get ready to discover the wonders of gravity as you conduct these fun and enlightening activities, all while gaining insights into the fundamental principles that govern our universe.

## 8.1 Hands-on learning to make the concepts come alive

Encouraging hands-on learning is a pivotal strategy for making scientific concepts, such as those related to gravity, come alive and resonate deeply with students. This approach capitalizes on the innate curiosity and exploratory nature of learners, creating an environment where they can actively engage with the subject matter and construct their understanding.

By offering hands-on activities and experiments, educators empower students to become scientists themselves, fostering a sense of ownership over their learning. These experiences allow students to see and feel the principles of gravity in action, transforming abstract ideas into tangible, memorable lessons.

Moreover, hands-on learning promotes critical thinking and problem-solving skills. Students not only witness the effects of gravity but also have the opportunity to analyze data, make predictions, and draw conclusions based on their observations. This process encourages a deeper level of comprehension, as students grapple with the forces and phenomena they encounter.

Furthermore, hands-on experiences transcend the classroom, making scientific concepts relevant to everyday life. Students can relate their newfound understanding of gravity to common occurrences, such as dropping an object or watching a pendulum swing. This connection bridges the gap between theory and reality, making learning meaningful and practical.

Lastly, hands-on learning fosters a lifelong love of science. By igniting curiosity and allowing students to explore the mysteries of the universe firsthand, educators lay the foundation for future scientific inquiry and discovery. Encouraging students to ask questions, seek answers, and explore the world around them through hands-on activities is an investment in their education and their future as informed, curious, and engaged citizens.

## 8.2 Simple gravity-related experiments readers can conduct themselves

**1. The Pendulum Swing:**

- Create a simple pendulum using a string or a piece of thread and a small object like a washer or a small ball.
- Attach one end of the string to a fixed point, like a doorknob, and let the object hang freely.
- Pull the object to one side and release it. Observe how it swings back and forth.
- You can change the length of the string or the weight of the object to see how it affects the pendulum's motion.

**2. Gravity and Falling Objects:**

- Find different objects of various sizes and weights, such as a feather, a coin, and a small ball.

- Hold them at the same height and drop them simultaneously. Observe which one hits the ground first.
- Discuss how gravity affects objects of different masses and shapes.

## 3. Balancing Act:

- Try to balance a pencil or a ruler on the edge of a table. It may seem challenging due to gravity.
- Experiment with different lengths and weights to find the right balance point.
- This demonstrates the force of gravity and the importance of balance.

## 4. Water Drops on a Coin:

- Place a clean, dry coin on a flat surface.
- Slowly drip water onto the coin, drop by drop.
- Observe how the water droplets behave on the surface of the coin due to surface tension and gravity.

## 5. Paper Airplanes:

- Create paper airplanes of different sizes and designs.
- Test their flight by throwing them gently into the air.
- Discuss how gravity affects the flight path and descent of the airplanes.

## 6. Straw Rocket Launch:

- Take a plastic drinking straw and a strip of paper.
- Roll the paper tightly around the straw, leaving a pointed end.
- Blow into the straw to launch the paper rocket.
- Observe how gravity pulls the rocket back down to the ground.

**7. Make Your Own Gravity Well:**

- Take a large, shallow container or tray and fill it with sand or flour.
- Place a heavy object, like a ball, in the center to represent a celestial body.
- Drop smaller objects (representing satellites or asteroids) onto the "gravity well" and observe how they move around the central mass, creating craters and impact patterns.

**8. Balloon Rocket:**

- Attach a piece of string or fishing line across a room.
- Inflate a balloon and tape it to a straw.
- Thread the string through the straw and hold one end.
- Let go of the balloon, and it will zoom along the string. Discuss how the escaping air propels the balloon forward, demonstrating action and reaction forces.

**9. Falling Water Drops:**

- Fill a small container with water, like a cup or a glass.
- Hold it over a sink or a bathtub and slowly pour the water out in a steady stream.
- Observe how the water flows downward due to gravity, forming droplets.
- Experiment by pouring water at different rates to see how it affects droplet size and speed.

**10. Paper Clip Float:**

- Fill a small bowl or dish with water.
- Carefully place a paperclip on the surface of the water. Note that it sinks.
- Using a tissue or a piece of wax paper, gently place it on the water's surface.
- Watch as the paperclip appears to "float" due to the surface tension of the water, defying gravity.

## 11. Magnetic Gravity:

- Take a small magnet and a few lightweight metal objects, such as paperclips or small screws.
- Hold the magnet above the metal objects and observe how they are attracted to it.
- Discuss how magnetism and gravity interact, and how they are different forces acting on objects.

## 12. Gravity-Powered Marble Run:

- Build a simple marble run using cardboard tubes, paper, and tape.
- Set up the marble run at an incline or on a staircase.
- Release a marble at the top, and observe how gravity propels it through the course.
- Experiment with different designs to see how they affect the marble's speed and path.

## 13. Homemade Gravity Well:

- Take a large, shallow container like a baking dish or a tray and fill it with sand or flour.
- Place a heavy object in the center to represent a celestial body (e.g., a large ball).
- Roll smaller objects (representing satellites or asteroids) onto the "gravity well" and observe how they move around the central mass, creating craters and impact patterns. This helps visualize how gravity affects the motion of objects in space.

## 14. Gravity and Falling Paper:

- Take a sheet of paper and crumple it into a ball.
- Hold the paper ball above a trash bin and release it.
- Observe how gravity pulls the paper ball toward the ground and discuss how the shape and weight of objects affect how they fall.

### 15. Gravity-Powered Water Fountain:

- Fill a plastic bottle with water and tightly seal it.
- Turn the bottle upside down and quickly remove your hand from the top.
- Water will flow out of the bottle, creating a miniature water fountain.
- Discuss how gravity pulls the water downward, creating the fountain effect, and how the speed of the flow depends on the size of the opening.

### 16. Gravity and Magnetic Attraction:

- Attach a strong magnet to the bottom of a container (e.g., a plastic cup).
- Place small metal objects like paperclips, screws, or coins on a table.
- Hold the container over the metal objects without touching them.
- Observe how gravity and magnetic attraction work together as the metal objects are attracted to the magnet inside the container.

### 17. Gravity and Air Resistance:

- Take a piece of paper and a small object (e.g., a coin).
- Hold the paper vertically with the object at the top.
- Release both the paper and the object simultaneously and observe their descent.
- Discuss how gravity affects the object's fall and how air resistance can slow it down.

### 18. Galileo's Leaning Tower Experiment:

- This experiment is inspired by Galileo Galilei's famous observation of falling objects from the Leaning Tower of Pisa.
- Find two objects of different masses, such as a feather and a coin.
- Hold them at the same height and drop them simultaneously.
- Observe how gravity affects objects of different masses and how they fall at the same rate in the absence of air resistance.

**19. Gravity and Elastic Potential Energy:**

- Take a rubber band or a stretchy hairband.
- Attach a small object (like a paperclip) to the band.
- Hold the band vertically and stretch it downward.
- Release the band, and the object will launch into the air.
- Discuss how gravity converts the stored elastic potential energy into kinetic energy, propelling the object upward.

**20. Magnetic Gravity Maze:**

- Create a maze using a piece of cardboard and magnets.
- Place a small metal ball (representing an object influenced by gravity) at the start of the maze.
- Use a magnet underneath the maze to guide the metal ball through the maze's twists and turns.
- Explore how magnetism and gravity can interact to control the movement of objects.

**21. Gravity and Coin Spin:**

- Find a flat, smooth surface like a tabletop.
- Take a coin and give it a gentle spin on the surface.
- Observe how the coin's spinning motion gradually slows down and stops due to the effects of friction and gravity.

**22. Gravity and Paper Helicopters:**

- Create a paper helicopter by cutting a sheet of paper into a specific shape with a vertical strip and rotor blades.
- Drop the paper helicopter from a height and observe how it spirals downward.
- Discuss how gravity affects the helicopter's descent and how air resistance plays a role in its spinning motion.

### 23. Egg Drop Challenge:

- Gather materials like eggs, various containers (e.g., cardboard boxes, plastic bottles), and cushioning materials (e.g., foam, bubble wrap).
- Design and build protective containers for the eggs to see if they can survive a fall from a certain height.
- Drop each container with an egg from the same height and observe which design best protects the egg from the force of gravity.

### 24. Center of Mass Exploration:

- Take a cardboard or foam sheet and cut out various shapes.
- Experiment with balancing these shapes on your fingertip or the edge of a table.
- Explore how the center of mass affects the stability of objects and how moving it can change the way objects balance.

### 25. Gravity and Marble Runs:

- Create a marble run using cardboard tubes, tape, and other materials.
- Experiment with different inclines and angles to observe how gravity affects the marble's speed and trajectory.
- Discuss concepts like potential and kinetic energy as the marble moves through the course.

### 26. Falling Domino Chain Reaction:

- Set up a line of dominoes close together, so they are nearly touching.
- Tip over the first domino, and observe how the gravitational force causes a chain reaction as each domino falls in succession.
- Explore how the initial push creates a gravitational domino effect.

## 27. Gravity and Water Pressure:

- Fill a plastic bottle with water and seal it tightly.
- Squeeze the sides of the bottle gently while observing the water level inside.
- Discuss how gravity influences the water pressure inside the bottle and why it changes when you squeeze it.

## 28. Gravity and Floating Objects:

- Fill a container with water and gather various objects made of different materials (e.g., plastic, wood, metal).
- Drop these objects into the water and observe which ones float and which ones sink.
- Explore how gravity interacts with buoyancy to determine whether objects rise or fall in water.

## 29. Gravity and Magnetic Marble Maze:

- Create a maze using a shallow cardboard box and magnets.
- Place a metal or magnetic marble at the start of the maze.
- Use magnets underneath the maze to guide the marble through the maze's twists and turns.
- Discuss how magnetism and gravity interact to control the marble's movement.

## 30. Gravity and Air Resistance (Parachute Drop):

- Create a small parachute by attaching a plastic bag or a square piece of cloth to a small object like a paperclip.
- Find a safe location with some height, like a staircase or balcony.
- Drop the parachute and observe how it falls slowly due to air resistance countering the force of gravity. Discuss how surface area affects the rate of descent.

### 31. Gravity and Magnetic Attraction (Magnet Drop):

- Drop a small magnetic object (like a magnetized washer or paperclip) down a transparent tube.
- Place strong magnets near the tube and observe how the falling object is affected by both gravity and magnetic attraction. Discuss the interplay between these forces.

### 32. Gravity and Time (Pendulum Clock):

- Construct a simple pendulum clock using a string, a weight (like a small bag of sand or beans), and a support structure.
- Adjust the length of the string and release the pendulum to observe its regular back-and-forth motion.
- Discuss how the pendulum's period (the time it takes to complete one swing) remains constant due to gravity, making pendulum clocks a precise timekeeping mechanism.

### 33. Gravity and Energy Transfer (Bouncing Balls):

- Take two balls of different materials, such as a rubber ball and a tennis ball.
- Drop them from the same height onto a hard surface and observe how they bounce differently.
- Discuss how the conservation of energy plays a role in how high the balls bounce and how gravity influences this process.

### 34. Gravity and Falling Liquids (Water and Air Pressure):

- Fill a drinking glass to the brim with water.
- Place a piece of cardboard or plastic over the mouth of the glass.
- Quickly invert the glass and note how the water remains suspended inside the glass due to air pressure. Gravity keeps the water from falling out.

## 35. Gravity and Sliding Objects (Friction Slide):

- Create a simple slide using a wooden board or a smooth piece of cardboard.
- Release objects of different shapes and sizes at the top of the slide and observe how they accelerate down the slope due to gravity and encounter friction.
- Discuss how the interaction between gravity and friction affects the objects' motion.

## 36. Gravity and Paper Bridge:

- Use sheets of paper or index cards to construct a bridge between two books or stacks of objects.
- Test the bridge's stability by placing lightweight objects (e.g., coins or small toys) on it.
- Discuss how gravity exerts force on the bridge and how its design influences its strength.

## 37. Gravity and Marble Roller Coaster:

- Build a miniature roller coaster using cardboard, paper, and tape.
- Create loops, drops, and twists for marbles to navigate.
- Release marbles at the highest point and observe how gravity propels them through the roller coaster's twists and turns.

## 38. Gravity and Surface Tension (Walking on Water):

- Fill a shallow container with water.
- Carefully place a paperclip on the water's surface, and it will float.
- Observe how surface tension, due to water molecules being attracted to each other, allows the paperclip to float without sinking.

### 39. Gravity and Magnetic Pendulum:

- Attach a small magnet to the end of a string.
- Hold a larger magnet beneath it, allowing the magnet on the string to swing freely.
- Observe how gravity and magnetic attraction affect the pendulum's motion, creating interesting patterns.

### 40. Gravity and Pendulum Art:

- Set up a pendulum using a weighted string or a small object hanging from a string.
- Place a canvas or paper underneath the pendulum's path.
- Release the pendulum, and it will create unique patterns as it swings due to the combined forces of gravity and inertia.

### 41. Gravity and Magnetic Levitation (Maglev Train):

- Create a simple "maglev" train using magnets.
- Place magnets on a track made of magnetic materials (like a metal ruler).
- Observe how the magnetic force opposes gravity, allowing the train to levitate and move along the track without friction.

### 42. Gravity and Pulley System:

- Build a simple pulley system using a string, a small bucket, and a hook.
- Use the pulley to lift small objects and discuss how the system allows you to overcome the force of gravity with less effort.

### 43. Gravity and Tilted Surfaces (Rolling Marbles):

- Set up inclined planes using books or cardboard.

- Release marbles at the top of the incline and observe how gravity causes them to roll down.
- Experiment with different incline angles to see how it affects the speed of descent.

## 44. Gravity and Helium Balloons:

- Fill two balloons, one with helium and the other with regular air.
- Observe how the helium balloon rises in the air while the air-filled balloon remains on the ground.
- Discuss how gravity pulls objects with mass towards the center of the Earth, and how helium's buoyant force counters gravity.

## 45. Gravity and DIY Water Clock:

- Create a simple water clock using two plastic bottles, a small hole in one bottle, and water.
- Let water drip from the first bottle into the second one and measure time intervals.
- Discuss how gravity pulls the water down to create a constant flow and how this flow can be used to measure time.

## 46. Gravity and Spinning Tops:

- Spin various spinning tops on a smooth surface.
- Observe how gravity, friction, and the top's shape affect its stability and spin duration.
- Discuss the balance between gravity and angular momentum in the spinning top's motion.

## 47. Gravity and DIY Marble Maze:

- Create a maze using cardboard, straws, and tape.
- Release marbles at the maze's entrance and navigate them through the maze.
- Discuss how gravity influences the marbles' paths as they roll through the maze.

### 48. Gravity and Jumping Pencils:

- Take two pencils and place them on a table so that one end of each pencil hangs over the edge.
- Place a small object, like a coin or eraser, on one of the pencil ends.
- Observe how the extra weight causes the pencil to pivot downward.
- Discuss how gravity and the lever effect contribute to this motion.

### 49. Gravity and Swinging Pendulum Art:

- Set up a pendulum using a weighted string and a canvas or paper underneath.
- Dip the pendulum's weight into paint and release it to create swinging pendulum art.
- Observe how gravity guides the pendulum's motion, resulting in unique artistic patterns.

### 50. Gravity and Water Clock with Multiple Containers:

- Create a multi-level water clock using several containers of varying sizes and small holes.
- Fill the containers with water, allowing them to slowly empty into one another.
- Observe how gravity controls the flow of water from one container to the next, creating a cascading timekeeping system.

## 8.3 Gravity Art Activity: Painting with Gravity

The "Gravity Art Activity: Painting with Gravity" is a hands-on creative project designed to help children and teenagers explore the concept of gravity in a fun and artistic way. Here's a detailed explanation of how this activity works:

**Materials Needed:**

- Large Sheets of Paper or Canvases: These will serve as the canvas for the artwork.
- A Variety of Non-Toxic Paint Colors: Use a range of colors to allow for creativity.
- Paintbrushes: While not for traditional painting, brushes can be used for mixing and adding fine details.
- Small Containers for Holding Paint: These containers will hold the paint that participants will use.
- Plastic Squeeze Bottles: These bottles will be used to apply the paint to the canvas.
- Easels or a Flat, Elevated Surface: Easels are ideal, but an inclined board can work too. These will hold the canvas at an angle, allowing gravity to play a role.

**Instructions:**

- *Setting Up:* Prepare a designated art space with easels or an inclined board where the painting will take place. Lay out all the materials.

- *Preparing the Paint:* Pour different paint colors into small containers or squeeze bottles. Make sure the paints are non-toxic and suitable for the age group participating.

- *Choosing Canvases:* Invite the children and teenagers to choose their canvas (large sheet of paper or canvas board), place it on the easel or inclined board, and select their paint colors.

- *Explaining Gravity's Role:* Explain that in this activity, they'll explore how gravity affects the movement of paint on their canvas. Gravity will play a role in determining how the paint flows and interacts with other colors.

- *Starting to Paint:* Encourage participants to start painting, but with a twist: instead of using brushes directly on the canvas, they will pour, drip, or squirt the paint onto the canvas. The goal is to experiment with gravity's influence on the paint.

- *Experimentation:* Participants can try different techniques. For example, they can drip paint from above and watch how it falls, pour it along the canvas's edge and observe the flow, or tilt the canvas to let the paint move in specific directions.

- *Observation:* As they apply the paint, they should actively observe how gravity affects the paint's movement. They can notice how colors blend, how paint drips or flows due to gravity, and any interesting patterns that emerge.

- *Drying and Reflecting:* Allow the artwork to dry completely. Once dry, participants can reflect on their creations. Discuss with them how gravity played a role in shaping their art. Ask questions about their observations and what they learned about gravity through this creative process.

- *Naming the Artwork:* Encourage participants to name their artwork based on their experience and what they see in the abstract patterns created by gravity.

The **"Gravity Art Activity: Painting with Gravity"** exercise offers a unique and multidimensional approach to understanding the concept of gravity. Here's an expansion of how this activity encourages participants to actively engage with the effects of gravity in a playful and artistic way, helping them grasp the concept in a memorable manner:

*Hands-On Exploration:* This exercise places the participants in the driver's seat of their artistic journey. By allowing them to physically manipulate paint and observe its behavior, it transforms the abstract concept of gravity into a tangible and interactive experience. This hands-on exploration taps into their innate curiosity and desire to interact with the world around them.

*Observational Learning:* As participants pour, drip, and tilt their canvases, they become keen observers of the natural world. They notice how the paint responds to the relentless pull of gravity. This observational aspect of the activity encourages scientific thinking and curiosity. Participants begin to ask questions about why paint drips in certain ways, why colors mix as they do, and how the angle of the canvas affects the paint's flow.

# The "Gravity Art Activity: Painting with Gravity

*Creative Expression:* Artistic endeavors like this foster creativity and self-expression. Participants are free to experiment with different techniques, color combinations, and patterns. By blending the scientific with the artistic, they are encouraged to think outside the box and explore new possibilities. This creative aspect not only makes learning fun but also allows them to express their unique perspectives on gravity.

*Multisensory Engagement:* Painting with gravity engages multiple senses. Participants see the colors mix and flow, hear the gentle sound of paint dripping, and feel the texture of the canvas. This multisensory experience reinforces their understanding of gravity and makes the learning process more immersive.

*Memorable Connection:* The artwork created during this exercise serves as a tangible and memorable connection to the concept of gravity. When participants look at their finished pieces, they recall the process and the role gravity played in shaping their art. This association between art and science leaves a lasting impression and enhances their retention of the concept.

*Collaborative Learning:* If done in a group setting, this activity promotes collaborative learning. Participants can share their observations and techniques with each other, leading to discussions and shared discoveries. Collaboration enhances the learning experience by allowing them to see different perspectives on how gravity influences their art.

*Reflective Thinking:* After the art has dried, participants engage in reflective thinking. They consider the patterns, colors, and shapes created by gravity and discuss their observations. This reflective phase encourages critical thinking and helps participants articulate what they've learned about gravity through their artistic experience.

*Sense of Ownership:* Naming their artwork based on their experience gives participants a sense of ownership and pride in their creations. It emphasizes that they are active participants in their learning journey, reinforcing their connection to the concept of gravity.

**Benefits:**

- Creative Learning: This activity combines artistic expression with scientific exploration, making learning about gravity a creative and engaging experience.
- Fostering Creativity: Participants have the freedom to experiment with different painting techniques, fostering creativity and imagination.
- Hands-on Understanding: By physically manipulating the paint and observing its behavior, children and teenagers gain a tangible understanding of how gravity influences real-world processes.
- Sense of Achievement: Completing their artwork provides a sense of accomplishment and pride in their creative and scientific abilities.

The **"Painting with Gravity"** activity is an excellent way to make the concept of gravity come alive, turning it into a memorable and enjoyable learning experience.

In summary, the **"Painting with Gravity"** exercise not only transforms a complex scientific concept into an accessible and enjoyable activity but also fosters creativity, curiosity, and memorable learning. By combining art and science, participants gain a deeper and more holistic understanding of gravity that extends beyond textbooks and lectures. This approach makes science come alive and encourages a lifelong appreciation for the wonders of the natural world.

## 8.4 Balloon Rocket Experiment

*Objective:* This simple yet engaging experiment helps children and teenagers understand the concept of action and reaction, which is a fundamental principle in physics, while also introducing them to the role of gravity in motion.

**Materials Needed:**

- A long piece of string (about 10-15 feet)
- A drinking straw
- Balloons (at least two)
- Tape
- Scissors
- Marker

**Instructions:**

➢ **Set Up the Experiment:**

- Cut a piece of string about 10-15 feet long and tie it between two fixed points in a room or outside (e.g., between two chairs or trees). Ensure the string is tight and level.
- Take a drinking straw and thread it onto the string. The straw should move freely along the string.

> **Prepare the Balloons:**

- Blow up two balloons and tie them shut.
- Use a marker to label one balloon "A" and the other "B."

## Attach the Balloons:

- Tape balloon "A" to one end of the straw, and balloon "B" to the other end. Make sure they are securely attached.

## Launch the Balloon Rockets:

- Stand in the middle of the string and hold the straw so that it's taut.
- Now, release the balloons simultaneously by letting them go. Watch as they move along the string.

## Observe and Experiment:

- Observe which balloon travels faster and why. Discuss the results with the participants.
- Try different combinations of balloons. For example, you can use two balloons of the same size or vary the size of the balloons while keeping one constant.
- Experiment by blowing up one balloon more than the other and see how it affects the race.

# Balloon Rocket Experiment

**Discussion Points:**

- Explain that the movement of the balloons is due to Newton's third law of motion, which states that "for every action, there is an equal and opposite reaction." In this case, the air rushing out of the balloons in one direction propels the balloons in the opposite direction.
- Discuss how gravity plays a role in this experiment by pulling the balloons downward and affecting their motion.
- Emphasize that the experiment showcases the interplay between different forces, including the force of the air escaping from the balloons and the force of gravity.

This activity not only provides a hands-on demonstration of fundamental physics principles but also encourages participants to make observations, ask questions, and experiment with different variables. It's a fun and interactive way to explore gravity's influence on motion and understand the concept of action and reaction. By rolling the marble down the inclined plane, participants are essentially creating their own mini roller coasters. They can vary the height and angle of the ramp to see how these factors affect the marble's speed and trajectory. This hands-on experimentation fosters a deeper understanding of gravity's role in the real world.

Moreover, as participants make predictions about how changing the ramp's characteristics will impact the marble's behavior, they are engaging in the scientific method—forming hypotheses, conducting experiments, and drawing conclusions. This not only enhances their grasp of gravity but also promotes critical thinking skills. It's a dynamic way to reinforce the idea that science isn't just about memorizing facts; it's about exploration and discovery.

This activity is also versatile and can be adapted to different age groups and learning levels. For younger children, it's a playful introduction to the concept of gravity and motion. For older students, it can serve as a more in-depth exploration of physics concepts and even lead to discussions about energy conservation, friction, and the mathematics of motion equations. Ultimately, it's a hands-on adventure that demonstrates that science is all around us, waiting to be explored and understood.

## 8.5 Paper Clip Divers: Exploring Buoyancy and Gravity

*Objective:* To demonstrate how buoyancy and gravity interact and influence the behavior of objects in water.

**Materials You'll Need:**

- Several paper clips (various sizes and shapes)
- A transparent plastic or paper cup
- An eyedropper
- Clean water
- Optional: A marker for decorating your paper clip divers

**Procedure:**

### Step 1: Crafting Your Paper Clip Diver

- *Select a paper clip:* Begin by choosing a paper clip to serve as your diver. Different paper clips may yield varying results, so feel free to experiment with different sizes and shapes. Some paper clips may be shaped like a "U," while others may be more like a "V." You can even use pliers to reshape them if desired.

- *Decorate (optional):* For added fun, consider decorating your paper clip diver with markers. You can give it a face, draw a tiny wetsuit, or let your creativity run wild. This step encourages artistic expression and personalization.

- *Bend your diver:* Carefully bend the paper clip into a shape resembling a diver. Create "arms" and "legs" by bending sections outward and upward. The diver should have an overall humanoid appearance.

### Step 2: Setting Up the Experiment

- *Prepare the cup:* Take your transparent plastic or paper cup and fill it with water. Leave some space at the top to ensure there's room for the water level to rise as you add more drops later in the experiment.

### Step 3: Observing Buoyancy and Gravity in Action

- *Deploy the paper clip diver:* Gently place your paper clip diver onto the water's surface within the cup. At this point, your diver should float on the water's surface. This initial buoyant force keeps it afloat because it is equal to the gravitational force pulling it downward.

- *Add drops of water:* Now, here comes the exciting part. Use the eyedropper to add water drops into the cup. Do this one drop at a time, and take note of the diver's behavior as you go. With each added drop, the cup becomes heavier due to the increased water weight.

**Step 4: Understanding the Results**

- *Observation:* Continue adding water drops one by one while observing the paper clip diver. At some point, you'll notice a change: your diver will slowly begin to sink.

- *Explanation:* This happens because as you add more water, the cup's weight increases, leading to a stronger gravitational force acting on the diver. The diver eventually becomes denser than the surrounding water, causing it to sink. It's a direct illustration of how the balance between buoyant force and gravity affects the behavior of objects in water.

**Step 5: Experimentation and Exploration**

- *Experiment further:* Encourage participants to experiment with different paper clip divers, cups of varying sizes, and the amount of water they add. They can predict when a diver will sink or float based on their observations and experiment to test those predictions.

**Key Concepts:**

- *Buoyancy:* Initially, the diver floats because the buoyant force, which pushes upward, equals the gravitational force, which pulls downward.

- *Gravity:* As you add more water, the gravitational force increases due to the added weight, causing the diver to sink.

- *Density:* The experiment demonstrates how an object's density relative to its surroundings determines whether it sinks or floats.

# Paper Clip Divers:

# Exploring Buoyancy and Gravity

The "Paper Clip Divers" activity offers an engaging and tangible way for children and teenagers to deepen their comprehension of fundamental physics principles, specifically those related to buoyancy and gravity. Rather than passively absorbing information, participants actively interact with the concepts, which can lead to more profound and lasting insights.

**Promoting Critical Thinking:**

One of the core benefits of this activity is its ability to foster critical thinking skills. As participants observe the paper clip diver's behavior, they are prompted to form hypotheses and make predictions. Why does the diver float initially, and why does it eventually sink? This encourages them to think critically about the forces at play, make connections between cause and effect, and develop problem-solving skills as they explore these questions.

**Encouraging Experimentation and Exploration:**

The "Paper Clip Divers" activity encourages experimentation, a fundamental aspect of scientific inquiry. Participants are encouraged to try different approaches, such as using various paper clip shapes or adjusting the water level in the cup. Through these experiments, they can test their hypotheses and refine their understanding. This hands-on exploration empowers young minds to ask questions, seek answers, and learn through trial and error.

**Sparking the Joy of Discovery:**

Learning through discovery can be incredibly motivating and enjoyable. When participants witness the paper clip diver's behavior change in response to their actions, it sparks a sense of wonder and curiosity. It's akin to solving a puzzle or unlocking a mystery, and this sense of achievement can be a powerful motivator for further exploration and learning.

**Interactive Play for Effective Learning:**

Learning through play is a well-documented and effective educational approach, particularly for children and teenagers. This activity combines playfulness with scientific concepts, making learning enjoyable and memorable. It transforms abstract theories into tangible experiences, making physics accessible and relatable.

In conclusion, the "Paper Clip Divers" activity is not just about demonstrating scientific principles; it's about engaging young minds in a dynamic and interactive way. By encouraging critical thinking, experimentation, and the joy of discovery, it not only helps children and teenagers understand buoyancy and gravity but also cultivates essential skills and a lifelong love for exploring the world through scientific inquiry.

## 8.6 Egg Drop Challenge: Exploring Gravity through Engineering

**Objective:**

The primary goal of the Egg Drop Challenge is for participants to design a device that can prevent a raw egg from breaking when it hits the ground after being dropped from a predetermined height. This activity is not just about saving the egg but also about understanding how gravity influences the egg's descent and how engineering can mitigate the forces of impact.

**Materials:**

- Raw eggs (each participant/group will need one)
- A variety of materials for constructing the protective contraption (e.g., straws, paper, rubber bands, plastic cups, balloons, cotton balls, tape, foam, etc.)
- Measuring tape or a marked height from which to drop the eggs
- A safe, open area for conducting the experiment

**Procedure:**

- Begin by explaining the challenge to the participants. They are tasked with designing and building a protective structure around a raw egg to ensure it survives a drop from a specified height.

- Participants can work individually or in small groups, fostering collaboration and brainstorming.

- Provide a selection of materials from which participants can choose to construct their protective devices. Encourage them to think creatively and consider the physics of falling objects influenced by gravity.

- Discuss the principles of gravity with the participants, explaining how it pulls objects downward and accelerates them as they fall.

- Instruct the participants to construct their protective contraptions, making sure they incorporate shock-absorbing features and cushioning materials to minimize the impact force.

- Once the contraptions are ready, measure the specified height from which the eggs will be dropped. Ensure that all participants have their contraptions and eggs prepared.

- Participants take turns dropping their egg-containing contraptions from the predetermined height. A designated spotter can ensure that eggs are dropped consistently.

- After each drop, carefully examine the condition of the egg. Did it survive without cracking or breaking? If it did, the design was successful in mitigating the force of gravity's impact.

- Encourage participants to make adjustments and improvements to their designs based on their observations and results from previous drops.

- Conduct multiple rounds of drops, allowing participants to refine their contraptions with each iteration.

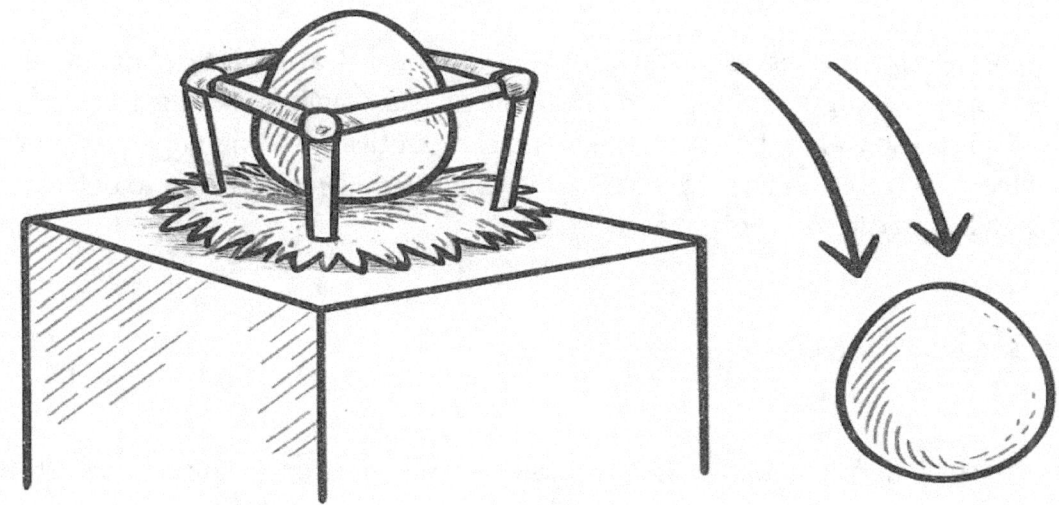

# Egg Drop Challenge: Exploring Gravity through Engineering

**Learning Outcomes:**

- Participants gain a hands-on understanding of how gravity influences the fall of objects.
- They explore engineering concepts and problem-solving skills to protect the egg.
- Creativity is encouraged as participants design unique contraptions.
- Participants learn through trial and error, refining their designs to achieve better results.
- The activity fosters teamwork and collaboration when conducted in groups.

The "Egg Drop Challenge" stands as an engaging, exhilarating, and educational experiment that skillfully translates the abstract concept of gravity into a tangible and unforgettable experience, tailored to capture the imagination of children and teenagers. Beyond its entertainment value, this challenge serves as a powerful educational tool, leaving a lasting impact by nurturing scientific curiosity and fostering crucial engineering skills.

**Engaging and Fun:**

The Egg Drop Challenge possesses an inherent allure that captivates participants of all ages. The prospect of designing a contraption that will save a fragile raw egg from gravity's relentless pull injects an element of excitement and competitiveness. This engagement factor ensures that learning becomes an enjoyable journey, motivating young minds to delve deeper into the underlying physics principles.

**Educational and Informative:**

At its core, the challenge is an exploration of gravity's influence on falling objects. Through hands-on experimentation, participants directly witness how gravity accelerates the egg during its descent and learn that, without intervention, the egg would succumb to the force of impact. This practical experience is a potent educational tool that imparts an intuitive understanding of gravity's effects, transcending the confines of traditional classroom instruction.

**Accessibility and Memorability:**

By translating the often abstract and distant concept of gravity into a tangible, real-world scenario, the Egg Drop Challenge makes it more accessible and relatable. Participants don't just hear about gravity; they witness its impact firsthand, leaving an indelible mark on their memory. This experiential approach enables children and teenagers to internalize and retain the fundamental principles of gravity with ease.

**Scientific and Engineering Skills:**

The challenge is a platform for budding scientists and engineers to shine. Participants are encouraged to employ critical thinking, creativity, and problem-solving skills to craft a contraption that defies gravity's destructive potential. They learn to identify factors influencing the egg's fall, such as velocity and impact force, and devise ingenious solutions to mitigate them.

**Building Lifelong Curiosity:**

Engaging in the Egg Drop Challenge nurtures a sense of wonder and inquisitiveness about the world around us. It ignites a spark of curiosity that extends far beyond the confines of the experiment itself. Participants are inspired to ask questions about the natural phenomena they encounter daily, ultimately fostering a lifelong love for science and discovery.

In sum, the "Egg Drop Challenge" transcends the boundaries of a simple experiment; it's a gateway to a deeper understanding of gravity and a testament to the power of experiential learning. It's an exciting journey that leaves young minds with an enduring appreciation for the invisible force that governs our world while equipping them with invaluable scientific and engineering skills for the future.

## 8.7 Balloon Rockets

*Objective:* The objective of the "Balloon Rockets" experiment is to demonstrate and understand Newton's third law of motion, which states that for every action, there is an equal and opposite reaction. Participants will also observe how the force of gravity influences the rocket's behavior by keeping it attached to the string before launch. This hands-on activity encourages critical thinking, problem-solving, and experimentation, while providing a fun and memorable way to learn about fundamental physics principles. Participants will have the opportunity to hypothesize, test, and draw conclusions about the factors that affect the rocket's performance, fostering a deeper understanding of scientific concepts.

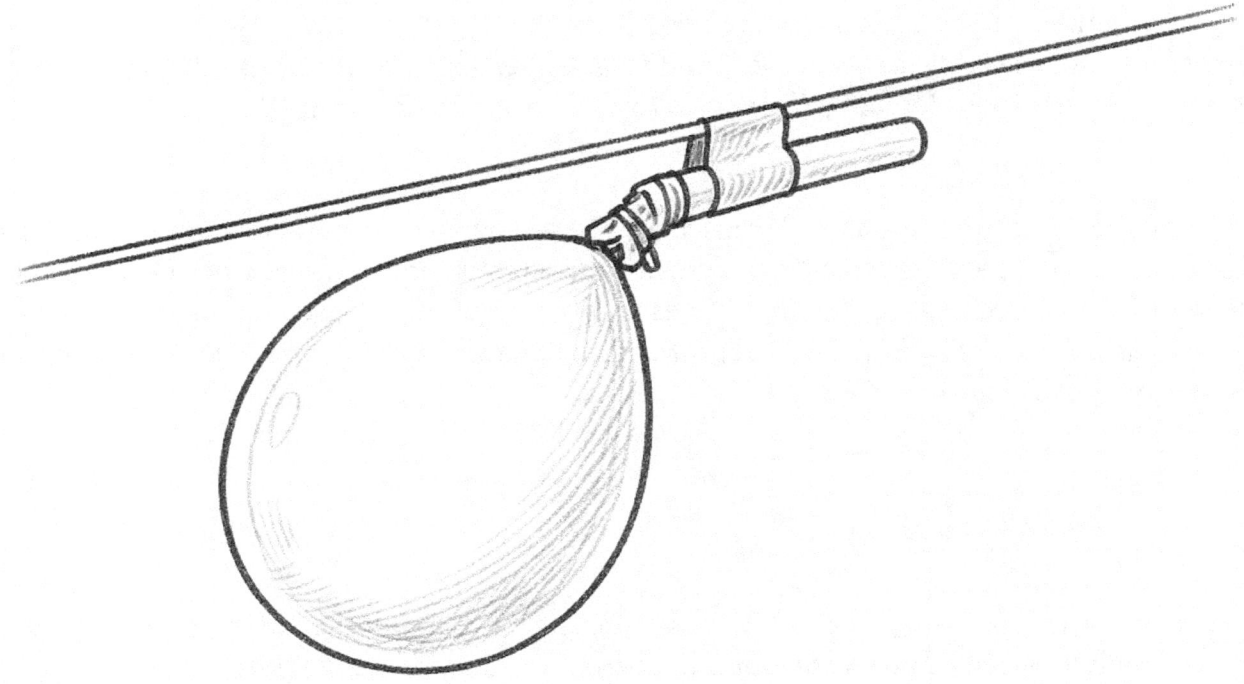

# Balloon Rockets

**Materials Needed:**

- String or a long piece of thread
- A plastic drinking straw (transparent ones work well)
- Balloons (several)
- Tape
- Scissors
- Marker (optional)

**Procedure:**

**Prepare the String:**

- Find a clear and open space, preferably indoors. A long hallway or a room with enough space to run the experiment is ideal.
- Cut a length of string or thread, making sure it spans the entire length of your chosen area. Securely attach the string horizontally at both ends, creating a tight line.

**Assemble the Rocket:**

- Take a plastic drinking straw and thread it onto the string. Ensure the straw can move freely along the string but isn't too loose.

**Prepare the Balloon:**

- Blow up a balloon but don't tie it.
- If you want to add a fun element, use a marker to decorate the balloon or draw a face on it.

**Attach the Balloon:**

- Carefully stretch the neck of the balloon over the end of the straw, ensuring it's secured tightly but not so tight that it can't be released later.

**Launch the Rocket:**

- With your rocket ready, stand at one end of the string.
- Hold the balloon, and keeping a firm grip on the straw, pull the balloon to the other end of the string.

- When you're ready for liftoff, let go of the balloon. The air escaping from the balloon propels the rocket along the string to the opposite end.

**Repeat and Experiment:**

- Try the experiment multiple times with variations. You can change the size of the balloon, the amount of air you blow into it, or the angle of the straw to see how these factors affect the rocket's performance.
- You can also challenge participants to predict which balloon size or angle will make the rocket travel the farthest.

**Learning Outcomes:**

The "Balloon Rockets" activity is a dynamic and engaging experiment that vividly demonstrates essential physics principles, including Newton's third law of motion and the ever-present force of gravity. This hands-on activity transcends traditional classroom learning by providing participants, especially children and teenagers, with a memorable and interactive way to grasp these fundamental concepts.

*Newton's Third Law of Motion:* At the heart of this activity lies Newton's third law, which states that for every action, there is an equal and opposite reaction. In the context of the "Balloon Rockets" experiment, participants witness this law in action as the release of air from the balloon propels the straw and the attached rocket forward. This striking visual representation of the law's application helps solidify the concept and fosters an intuitive understanding of how forces interact in the physical world.

*Force of Gravity:* While the primary focus is on Newton's third law, the "Balloon Rockets" activity also indirectly highlights the force of gravity. Although gravity is not directly responsible for the balloon's propulsion, it plays a crucial role in the experiment. Gravity keeps the rocket and the straw tethered to the string before the balloon is released. As the balloon expels air, the rocket accelerates due to the reaction force, moving in opposition to gravity's pull. This interplay between Newton's third law and gravity is a central theme of the activity.

*Interactivity and Engagement:* One of the strengths of the "Balloon Rockets" activity is its interactive nature. Participants actively engage with the experiment, blowing up balloons, taping them to straws, and then witnessing the exhilarating launch of their homemade rockets. This hands-on approach fosters a sense of excitement and curiosity, making learning about physics a thrilling adventure rather than a passive exercise.

*Memorable and Relatable:* By transforming abstract concepts into a tangible and relatable experience, the "Balloon Rockets" activity ensures that the lessons learned are memorable and enduring. Children and teenagers can vividly recall their balloon rocket experiments, making it easier to retain knowledge about Newton's third law and gravity.

*Critical Thinking and Problem-Solving:* The activity also encourages critical thinking and problem-solving skills. Participants must consider various factors, such as the size of the balloon, the force of air expulsion, and the angle of the straw, to optimize the rocket's performance. This element of experimentation cultivates important skills that extend beyond the realm of physics.

In summary, the "Balloon Rockets" activity is a captivating and instructive experiment that effectively conveys complex scientific principles in an accessible and enjoyable manner. It provides a platform for children and teenagers to explore the laws of motion and the force of gravity through hands-on engagement, leaving them with a lasting appreciation for the wonders of physics.

## 8.8 Gravity-Powered Cars experiment

Experiment Description: In the "Gravity-Powered Cars" experiment, participants are tasked with designing and constructing small cars that rely solely on the force of gravity to propel them downhill. This hands-on activity offers a creative and engaging way to explore the fundamental principles of gravity, motion, and design.

**Materials Needed:**

- Cardboard (for the car's body)
- Plastic bottles (for wheels)

- Rubber bands (for propulsion)
- Craft supplies (such as tape, glue, markers, and scissors)
- A ramp or an inclined surface (e.g., a sloped wooden board)
- Measuring tape or a ruler

**Procedure:**

- Begin by discussing the concept of gravity with the participants. Explain that gravity is the force that pulls objects toward the center of the Earth and that it plays a crucial role in the movement of objects.

- Challenge the participants to design and build their gravity-powered cars using the provided materials. They can use their creativity to come up with unique car designs, considering factors like the shape of the car, the size of the wheels, and the attachment of rubber bands for propulsion.

- Once the cars are constructed, set up an inclined surface or ramp. This will serve as the "race track" for the gravity-powered cars.

- Participants can experiment with different variables to optimize their car's performance. They can adjust the angle of the ramp, the tension of the rubber bands, or the weight distribution of the car to see how these factors affect the car's speed and distance traveled.

- Hold a friendly competition where each participant releases their car from the same starting point on the ramp, and the goal is to see which car travels the farthest or fastest purely due to gravity.

**Key Concepts Demonstrated:**

*Gravity:* Participants learn that gravity is the force that pulls objects toward the Earth's center. In this experiment, gravity is the sole source of propulsion for the cars as they roll downhill.

*Motion:* The experiment illustrates the principles of motion, including acceleration and velocity. Participants observe how gravity causes the cars to accelerate as they move downhill and how different design choices affect the cars' velocity.

*Energy Transfer:* Participants discover that the potential energy stored in the rubber bands is converted into kinetic energy as the cars move. This demonstrates the transfer of energy, a fundamental concept in physics.

*Design and Engineering:* Participants engage in the engineering design process by creating their gravity-powered cars. They learn that design choices, such as the shape of the car, the size of the wheels, and the tension of the rubber bands, can impact the car's performance.

*Experimental Inquiry:* Through experimentation and adjustments, participants practice the scientific method by making observations, forming hypotheses, and testing their ideas. They learn to analyze results and draw conclusions based on their experiments.

*Newton's Third Law:* Although not explicitly stated, participants indirectly observe Newton's third law of motion, which states that for every action, there is an equal and opposite reaction. As the air rushes out of the balloon, it propels the car forward in the opposite direction, showcasing this law in action.

# Gravity-Powered
# Cars Experiment

**Learning Outcomes:**

The objective of the "Gravity-Powered Cars" experiment is to introduce participants to the concept of gravity as a fundamental force that affects the motion of objects. Through hands-on construction and experimentation, participants will gain insights into how design choices impact the performance of gravity-powered vehicles. This activity promotes creativity, problem-solving, and scientific inquiry while making physics concepts tangible and enjoyable for children and teenagers.

**By engaging in this experiment, participants will:**

*Understand Gravity:* They will learn that gravity is a force that pulls objects toward the Earth's center, and it can be harnessed to create motion.

*Explore Motion:* Participants will observe the principles of motion, including acceleration and velocity, in action as their cars roll downhill.

*Apply Design and Engineering:* Through designing and building their cars, participants will apply engineering concepts and realize that design choices directly influence the car's performance.

*Practice the Scientific Method:* This activity encourages participants to follow the scientific method by making hypotheses, conducting experiments, analyzing results, and drawing conclusions.

*Foster Creativity:* Participants can exercise their creativity by coming up with unique car designs and exploring different materials and configurations.

*Enhance Problem-Solving Skills:* As they encounter challenges and make adjustments to improve their cars, participants will enhance their problem-solving skills.

*Enjoy Hands-On Learning:* The experiment offers a fun and interactive way to learn about gravity and motion, making science enjoyable and accessible.

These learning outcomes reflect the educational goals of the experiment, emphasizing both scientific understanding and practical skills related to gravity and engineering. Participants will not only learn about gravity but also apply their knowledge to build functional vehicles, fostering a deeper understanding of physics concepts.

The "Gravity-Powered Cars" experiment serves as an effective educational tool to demystify complex scientific concepts, spark curiosity, and inspire future scientists and engineers. It underscores the idea that science can be both fun and educational, encouraging a lifelong interest in learning.

## 8.9 Balancing Act

**Experiment Description:**

In the "Balancing Act" experiment, participants are presented with various objects and are challenged to balance them on different parts of their bodies, such as their fingertips, palms, or even their noses. The objective is to discover how the distribution of mass within an object affects its stability when placed in different positions.

**Objective:**

The primary objective of the "Balancing Act" experiment is to help participants grasp the concept of the center of gravity and its role in determining the stability of objects. By actively engaging with this hands-on activity, participants gain insights into the fundamental principle that an object is most stable when its center of gravity is positioned directly above the point of support.

**Materials Needed:**

- *Various Objects:* Collect a variety of objects with different shapes and sizes. These could include books, wooden blocks, toy figurines, or household items like a broomstick or a spoon.

- *Open Space:* Ensure you have enough space for participants to move around and balance the objects on different parts of their bodies.

- *Optional:* You can have a ruler or measuring tape on hand to measure the height at which participants successfully balance an object.

# BALANCING ACT

**Procedure:**

- *Introduction:* Begin by explaining the concept of the center of gravity to the participants. Describe how the center of gravity is the point within an object where the force of gravity appears to act. Emphasize that objects tend to be most stable when their center of gravity is directly above the point of support.

- *Balancing Objects:* Provide participants with a selection of objects and encourage them to experiment with balancing these objects on their fingertips, palms, or other body parts. They can start with simple and symmetrical objects before moving on to more irregular shapes.

- *Observation and Discussion:* As participants attempt to balance the objects, encourage them to make observations about which objects are easier or more challenging to balance. Discuss their findings and guide them in understanding how the distribution of mass within each object influences its balance.

- *Variations:* To deepen the experiment, participants can try balancing objects at different heights above the ground, such as on the tips of their fingers or close to their palms. This variation allows them to explore how the height of the center of gravity impacts stability.

**Learning Outcomes:**

The "Balancing Act" experiment provides several valuable learning outcomes:

- Participants gain a practical understanding of the concept of the center of gravity and its significance in determining the stability of objects.

- They develop observational and analytical skills as they assess which objects are easier or more challenging to balance.

- This hands-on activity fosters critical thinking as participants explore how changes in the distribution of mass affect an object's balance.

- Participants also enhance their fine motor skills and coordination through the physical act of balancing objects on their fingertips or other body parts.

The "Balancing Act" experiment offers a dynamic and approachable approach to acquainting participants with the concept of the center of gravity and its impact on the stability of objects. This experiment goes beyond conventional classroom learning, transforming scientific principles into a tangible, hands-on experience. Here are some key takeaways that emphasize the significance of this activity within the realm of science education for children and teenagers:

*Engagement:* The "Balancing Act" experiment captivates participants by turning abstract physics concepts into a captivating physical challenge. It encourages active participation and curiosity, motivating learners to explore and comprehend the science behind every day phenomena.

*Observational Skills:* Participants develop their observation and analysis abilities as they discern which objects are more or less challenging to balance. This fosters a scientific mindset, encouraging learners to pay attention to detail and form hypotheses about object stability.

*Critical Thinking:* The experiment prompts critical thinking as participants investigate how variations in mass distribution affect an object's equilibrium. By engaging in hands-on problem-solving, learners refine their ability to analyze situations and draw conclusions based on evidence.

*Practical Understanding:* Instead of relying solely on theoretical explanations, participants gain a practical understanding of the center of gravity and its real-world implications. This tangible experience enhances their grasp of fundamental physics principles.

*Fine Motor Skills:* Balancing objects on different parts of their bodies enhances participants' fine motor skills and coordination. This adds a physical dimension to the learning process, promoting overall development.

*Science Education Enhancement:* As a valuable addition to science education, the "Balancing Act" experiment transforms complex scientific ideas into an accessible and enjoyable learning opportunity. It promotes a deeper connection to physics concepts and fosters a lifelong interest in science.

The "Balancing Act" experiment is a versatile tool that bridges the gap between theory and practice, making science come alive for children and teenagers. By actively engaging with the concept of the center of gravity, participants not only expand their scientific knowledge but also cultivate essential skills in observation, critical thinking, and problem-solving. This experiment serves as a testament to the power of hands-on learning experiences in science education.

## 8.10 Galilean Moons Simulation

*Objective:* The objective of the Galilean Moons Simulation is to provide a hands-on, visual representation of the concept of gravity in space by simulating the orbits of Jupiter's four largest moons: Io, Europa, Ganymede, and Callisto. Participants will use different-sized balls or objects to represent the moons and experiment with their orbits around a central point. Through this experiment, learners will gain insights into how gravity influences celestial bodies' motions and understand the dynamics of orbital mechanics.

**Materials Needed:**

- *Four Balls or Objects:* These will represent Jupiter's four largest moons—Io, Europa, Ganymede, and Callisto. It's essential to use objects of varying sizes to demonstrate the moons' differing masses.

- *A Central Point:* This can be a sturdy support, such as a small table or a central marker on a flat surface, to serve as the gravitational center (representing Jupiter).

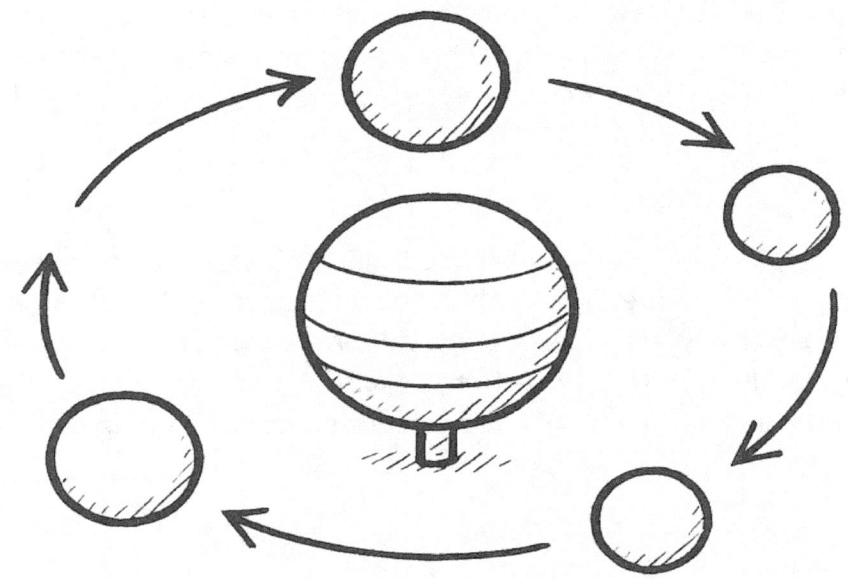

# Galilean Moons Simulation

- Space for Experimentation: Ensure you have a clear, open space where participants can set up the simulation without obstructions.

**Procedure:**

- *Setup:* Place the central point (representing Jupiter) on a flat surface. Make sure it is stable and won't move during the experiment.

- *Select Objects:* Designate four different-sized balls or objects to represent the Galilean moons. Label them as Io, Europa, Ganymede, and Callisto for clarity.

- *Orbiting the Central Point:* Have participants stand around the central point, each holding one of the moon objects. They should start by placing their objects at rest on the central point.

- *Initiating the Orbits:* Instruct participants to give their moon objects a gentle push, initiating their orbits around the central point (Jupiter). It's important to emphasize that the direction and speed of the initial push can affect the moons' orbits.

- *Observe and Experiment:* Encourage participants to observe the motion of the moon objects as they orbit the central point. They should take note of the differences in orbit size, speed, and behavior among the moons.

- *Discussion:* After the initial orbits, gather participants for a discussion. Explore concepts related to gravity in space, such as how the mass of each moon affects its orbit and how gravity keeps celestial bodies in stable orbits. Discuss why some moons might have larger or faster orbits than others.

**Key Concepts Demonstrated:**

*Gravity in Space:* The experiment vividly illustrates how gravity operates in space by showing how celestial bodies (moons) are drawn towards a central massive object (Jupiter) and continuously orbit it.

*Orbital Mechanics:* Participants learn about the basics of orbital mechanics, including how the mass and initial velocity of an object influence its orbit.

*Differences Among Moons:* The varying sizes and initial conditions of the moon objects highlight that different celestial bodies can have different orbits due to their mass and velocity.

*Stability of Orbits:* The experiment demonstrates how gravity maintains stable orbits, preventing celestial bodies from either crashing into the central object or flying away into space.

The Galilean Moons Simulation is a hands-on and interactive experiment designed to bring the concept of gravity and celestial motion to life. By simulating the orbits of Jupiter's four largest moons—Io, Europa, Ganymede, and Callisto—this activity bridges the gap between abstract space concepts and practical, observable phenomena. It offers several educational benefits and encourages active participation and learning:

*Engagement:* Participants are actively involved in setting up and conducting the experiment. They become "astronomers" for a brief moment, taking on the roles of celestial bodies and witnessing firsthand the effects of gravity on their orbits.

*Tangible Understanding:* The use of physical objects to represent celestial bodies makes complex astronomical concepts tangible and relatable. Participants can see, touch, and manipulate the objects, enhancing their comprehension of abstract ideas.

*Observation Skills:* The experiment prompts participants to carefully observe the behavior of the moon objects as they orbit the central point. This fosters the development of observational skills and the ability to draw conclusions from what they see.

*Critical Thinking:* Throughout the activity, participants are encouraged to think critically. They can experiment with different initial conditions, such as the strength of the push or the direction of the orbits, and observe how these factors influence the moons' paths.

*Discussion and Exploration:* After the simulation, a discussion provides an opportunity to delve deeper into concepts related to gravity, orbital mechanics, and the unique characteristics of each moon. Participants can ask questions, share their observations, and explore the reasons behind the differences in moon behavior.

*Practical Application:* The experiment lays the groundwork for understanding the gravitational relationships between celestial bodies, which is a fundamental concept in astronomy and astrophysics.

*Inspiration:* For aspiring astronomers and scientists, this activity can serve as an inspiring introduction to the wonders of space and the study of celestial motion.

The Galilean Moons Simulation is a valuable educational tool that sparks curiosity, fosters understanding, and encourages exploration in the realms of gravity and space science. It provides a memorable and engaging experience that can leave a lasting impression on children and teenagers interested in the mysteries of the cosmos.

# Conclusion

In the journey through the pages of "The Gravity Guide: Unveiling the Universe's Hidden Force," we have embarked on a captivating exploration of one of the most fundamental forces in the cosmos, gravity. We began by examining its omnipresent influence in our daily lives, from the simple act of dropping an object to the profound ways it shapes the cosmos. Through engaging narratives, experiments, and thought-provoking discussions, we unraveled the mysteries of gravity's role in celestial bodies, the fabric of spacetime, and the intriguing phenomena of black holes and dark matter.

As we delved into the scientific revelations and technological marvels that have emerged from our understanding of gravity, we explored how this force impacts our world in profound and often surprising ways. From the precise timekeeping of pendulum clocks to the awe-inspiring wonder of space exploration, we witnessed how humanity's mastery of gravity has transformed our perception of the universe and our place within it. Moreover, we marveled at the brilliant mind of Albert Einstein, whose theory of general relativity reshaped our cosmic understanding, paving the way for technologies like GPS and opening new frontiers in our exploration of space.

Throughout this journey, we have not only gained a comprehensive understanding of gravity's principles but also a profound appreciation for the beauty and complexity of the cosmos. We've learned that gravity is not merely a force that keeps our feet firmly planted on Earth; it is the unseen hand that guides the dance of the stars, the orchestrator of celestial movements, and the key to unraveling the secrets of the universe. As we conclude this exploration, may we carry forward a sense of wonder and curiosity, ever eager to continue our voyage of discovery through the limitless expanse of the cosmos, forever bound by gravity.

## Key takeaways

*"The Gravity Guide: Unveiling the Universe's Hidden Force"* has taken us on an exhilarating journey through the enigmatic world of gravity, revealing its profound influence on our lives and the cosmos. Here are the key takeaways and the practical and fascinating aspects of gravity:

1. **Universal Force:** Gravity is a universal force that affects all objects with mass. It's not just the reason why things fall to the ground; it's also responsible for keeping planets in orbit around the sun and stars clustered in galaxies.

2. **Equivalence Principle:** Einstein's theory of general relativity introduced the concept that gravity is not just a force but also the result of the curvature of spacetime by mass. This principle has profound implications for our understanding of gravity's effects on the universe.

3. **Practical Applications:** Understanding gravity has led to practical applications like GPS systems, where precise timing and knowledge of gravitational effects are crucial for accurate navigation. Gravity also plays a role in satellite orbits and space exploration.

4. **Cosmic Mysteries:** Gravity is intimately connected to some of the universe's most captivating mysteries, including black holes, dark matter, and dark energy. These phenomena continue to baffle scientists and expand our knowledge of the cosmos.

5. **Hands-On Learning:** Engaging in hands-on experiments and observations allows us to grasp the concepts of gravity more profoundly. Activities like pendulum experiments, marble runs, and simulations help make gravity's effects tangible and relatable.

6. **Interconnectedness:** Gravity is the force that binds celestial bodies, shapes galaxies, and governs the cosmic dance of the universe. It underscores the interconnectedness of all things and deepens our appreciation for the cosmos.

7. **Curiosity and Wonder:** Gravity fuels our innate curiosity about the universe. It inspires us to ask questions, seek answers, and explore the cosmos. Our journey through *"The Gravity*

*Guide"* has reinforced the idea that the quest for knowledge and the marvels of the universe are bound together by the force that shapes it all: Gravity.

## Hidden force that shapes our world and inspires curiosity

In conclusion, gravity is not just a force that keeps us grounded; it's a force that shapes our understanding of the universe and drives our exploration of the cosmos. By delving into the practical applications and intriguing phenomena of gravity, we unlock the secrets of our world and the broader universe, leaving us in perpetual awe of the hidden force that binds us to the cosmos.

As we conclude our journey through the captivating pages of "The Gravity Guide: Unveiling the Universe's Hidden Force," I invite you, dear readers, to take a moment to appreciate the profound and often invisible influence of gravity in our lives and the cosmos. In the depths of this exploration, we have unraveled the mysteries of gravity's pull, from the humblest of everyday experiences to the grandeur of celestial motions. But more than just understanding, let us pause to embrace the wonder and inspiration that gravity offers.

Consider for a moment the magic of a falling leaf, the graceful orbit of a moon around its planet, or the cosmic ballet of galaxies dancing through the cosmos, all guided by the silent hand of gravity. This force is a reminder that the world around us is a mosaic of interconnected phenomena, each thread woven by gravity's embrace. It's an invitation to see the world with fresh eyes, to question, to marvel, and to seek the hidden truths that underpin our existence.

As you gaze up at the night sky, knowing that gravity shapes the orbits of celestial bodies, or as you watch an object fall to the ground, aware of the universal force at play, let it inspire your curiosity. Let it be a catalyst for exploration, a reminder that the universe is a vast, enigmatic playground waiting to be explored. Embrace the questions that arise, for it is through these questions that we uncover the hidden marvels of the cosmos.

May this journey through the world of gravity spark your curiosity, deepen your appreciation for the forces that shape our world, and inspire you to continue exploring the mysteries of the universe. After all, in the dance of the cosmos, gravity is both the choreographer and the muse, beckoning us to join in the eternal quest for knowledge and wonder.

# Epilogue: The Future of Gravity

## Toward Gravity's Horizon

We bend the void with silent grace,
To ride the waves that stars embrace.
Through gravity's song, we learn to steer,
Past light's delay and time's frontier.
A whisper warped can now transmit,
Where matter folds, our engines flit.
No wheel, no wire—just spacetime's thread,
To build, to speak, where angels dread.
From black hole hearts, we mine the flame,
The cosmos bends, but calls our name.

What lies ahead when we peer beyond what we know?

Gravity, once simply the force that made apples fall and planets orbit, has evolved into something far more mysterious and profound. From Newton's falling fruit to Einstein's spacetime curvature, we've followed gravity through centuries of thought, discovery, and reimagination. Yet, for all that we've learned, we stand at the edge of a cosmic cliff, staring into a future filled with unanswered questions.

Will we one day manipulate gravity the way we now manipulate electricity? Could we neutralize it, reverse it, or perhaps even travel by it: through gravity waves, wormholes, or quantum tunnels? Might we find a unified theory that weaves gravity seamlessly into the fabric of quantum mechanics, a "Theory of Everything" that finally dissolves the division between the large and the small?

The pursuit of gravity's future isn't confined to chalkboards and observatories. It's in the clocks that correct themselves in GPS satellites, in the hunt for dark matter deep underground, and in the silent signals of merging black holes reaching us across billions of light-years. Each wave, each particle, each falling object whispers a secret, and we are just beginning to learn how to listen.

If history has shown us anything, it's that the greatest truths often begin with the simplest curiosities. A falling apple. A child asking, "Why do things fall down?" A spacecraft veering just slightly off course. These are not endpoints. They are invitations.

The future of gravity is not merely about forces and formulas. It is about imagination. About daring to ask questions no one has answered yet. It is about what happens when wonder meets knowledge, and the mind dares to leap—where nature, science, and soul collide, even under the rule of gravity.

So, here's to gravity — the unseen force that holds everything together, and yet may be the key to taking us beyond everything we know.

— *Pyura Anshuman*

# Appendix: Further Exploration

For those eager to embark on further exploration and delve deeper into the captivating realm of gravity, there are numerous avenues to satisfy your curiosity and expand your understanding:

1. **Read More:** Consider delving into books authored by renowned physicists and cosmologists, such as Stephen Hawking's "A Brief History of Time" or Brian Greene's "The Fabric of the Cosmos." These books provide insightful perspectives on gravity and its role in the universe.

2. **Online Resources:** The internet offers a wealth of educational resources, including online courses, lectures, and simulations related to gravity and astrophysics. Websites like NASA's official site and educational platforms like Coursera and Khan Academy provide accessible and informative content.

3. **Planetarium Visits:** Explore your local planetarium, where you can attend fascinating presentations, view celestial objects through powerful telescopes, and engage in interactive exhibits that showcase the wonders of the universe.

4. **Join a Science Club or Society:** Consider becoming part of a local or online science club or society. These groups often host discussions, lectures, and events related to gravity, space exploration, and astrophysics.

5. **Observe the Night Sky:** Invest in a telescope or binoculars and spend time stargazing. Observing celestial objects like planets, stars, and even the moon can deepen your connection to gravity's role in the cosmos.

6. **Experimentation:** Continue with hands-on experiments and activities related to gravity. Explore new experiments or revisit previous ones to deepen your understanding and share your discoveries with others.

7. **Document Your Observations:** Keep a journal or blog where you document your thoughts, questions, and observations related to gravity and the universe. Sharing your insights with others can foster meaningful discussions and connections.

8. **Connect with Experts:** Reach out to physicists, astronomers, and educators through social media or local events. Engaging with experts can provide valuable insights and answers to your questions.

9. **Attend Lectures and Workshops:** Seek out lectures, workshops, and conferences related to astrophysics and gravitational studies. These events often feature leading scientists discussing the latest discoveries in the field.

10. **Share Your Passion:** Encourage others to explore the wonders of gravity and the universe. Share your knowledge, enthusiasm, and resources with friends, family, or fellow enthusiasts, creating a community of like-minded individuals.

Remember that the journey of exploration is ongoing and ever-evolving. Gravity, as a fundamental force of the universe, offers an endless source of inspiration and discovery. Whether you choose to continue your studies as a casual enthusiast or a dedicated scholar, the pursuit of understanding gravity and the cosmos is a journey that promises continual fascination and wonder.

**List of accessible resources, videos, and websites for readers interested in learning more.**

Here's a list of accessible resources, videos, and websites that will help readers delve deeper into the fascinating world of gravity and astrophysics:

**Websites:**

*NASA's Gravity and Orbits:* NASA's website offers a wealth of resources on gravity, orbital mechanics, and space exploration. Explore articles, images, and interactive tools to learn more.

*Khan Academy - Physics and Astronomy:* Khan Academy offers free, comprehensive lessons on physics and astronomy topics, including gravity and its role in the cosmos.

*PBS NOVA - Secrets of the Universe:* PBS NOVA provides a series of engaging videos and articles on topics related to astrophysics, black holes, and the mysteries of the universe.

*European Space Agency (ESA) - Education:* ESA's education portal offers a range of educational materials, videos, and interactive tools for students and curious learners interested in space and gravity.

**YouTube Channels:**

*PBS Space Time:* This YouTube channel explores complex topics in astrophysics, including gravity, black holes, and the nature of the universe. It's presented in an engaging and accessible manner.

*Fraser Cain's Universe Today:* Fraser Cain provides informative videos on various aspects of space and astronomy, including gravity's role in shaping the cosmos.

*Vsauce:* Vsauce's videos delve into intriguing scientific questions, often touching on concepts related to gravity and physics. The channel aims to make complex ideas understandable.

**Online Courses:**

*Coursera - Astrophysics Courses:* Coursera offers a range of astrophysics courses from top universities and institutions, many of which cover gravity and its effects in space.

*edX - Astronomy and Space Science:* Explore free courses on astronomy and space science, where you can learn about gravity's role in celestial mechanics.

**Books:**

*"A Brief History of Time" by Stephen Hawking:* This bestseller offers a clear and accessible exploration of complex astrophysical concepts, including gravity.

*"The Fabric of the Cosmos" by Brian Greene:* Brian Greene's book provides insights into the nature of the universe, including discussions on gravity and its role in spacetime.

These resources cater to a wide range of interests and levels of expertise, making them accessible to both beginners and those seeking in-depth knowledge about gravity and its impact on the universe. Whether you prefer articles, videos, or interactive courses, there are plenty of avenues to explore the wonders of gravity and astrophysics.

# Gravity in 10 Milestones

A Timeline of Humanity's Journey with the Hidden Force

## 1. ~400 BCE – Aristotle's Natural Motion
Gravity is imagined as a natural tendency of heavy objects to seek the center of the universe (Earth).

## 2. ~1600 – Galileo's Experiments
Galileo Galilei drops spheres from the Leaning Tower of Pisa, showing objects fall at the same rate regardless of mass.

## 3. 1687 – Newton's Universal Law of Gravitation
Isaac Newton publishes Principia Mathematica, defining gravity as a force acting between all masses in the universe.

## 4. 1789 – Cavendish Measures Gravity
Henry Cavendish determines the gravitational constant (G) using a torsion balance, "weighing the Earth" for the first time.

## 5. 1915 – Einstein's General Relativity
Albert Einstein proposes gravity is not a force, but the warping of spacetime by mass and energy.

## 6. 1919 – Solar Eclipse Confirms Relativity

Arthur Eddington observes starlight bending around the Sun, experimentally verifying Einstein's predictions.

## 7. 1969 – Gravity on the Moon

Apollo 11 astronauts experience lower gravity, showcasing the universality of gravitational pull beyond Earth.

## 8. 1974–1990s – Binary Pulsars and Relativity

Observations of neutron star pairs indirectly confirm gravitational wave emission, matching Einstein's predictions.

## 9. 2015 – First Detection of Gravitational Waves (LIGO)

Humanity "hears" the universe as LIGO detects spacetime ripples from merging black holes.

## 10. Future – Gravity as a Tool for Space and Time

From gravitational propulsion and communication to spacetime engineering, gravity could become a technology of exploration.